T0168025

Discovering
October Roads

Discovering October Roads

Fall Colors and Geology in Rural East Tennessee

Harry Moore and Fred Brown

The University of Tennessee Press
KNOXVILLE

First Edition.

All photographs and drawings by Harry Moore.

The paper used in this book meets the minimum requirements of ANSI/NISO Z39.48-1992 (R 1997) (Permanence of Paper). The binding materials have been chosen for strength and durability.

Library of Congress Cataloging-in-Publication Data

Moore, Harry, 1949–
Discovering October roads: fall colors and geology in rural east Tennessee / Harry Moore and Fred Brown.
p. cm.
Includes bibliographical references and index.
ISBN 1-57233-122-4 (cl.: alk. paper)
ISBN 1-57233-123-2 (pbk.: alk. paper)
1. Tennessee, East—Guidebooks. 2. Natural history—Tennessee, East—Guidebooks. 3. Geology—Tennessee, East—Guidebooks.
4. Historic sites—Tennessee, East—Guidebooks. I. Brown, Fred. II. Title.
F442.1 .M66 2001
917.68'80454—dc21 00-011820

For R. E. McLaughlin
and Eddie Barker

Contents

Illustrations

Plates

Figures

Maps

Table

Acknowledgments

The preparation of a manuscript covering the topics of fall color, cultural history, and geology required the assistance of several key persons, whom we would like to acknowledge here. Our understanding of the fall color season and the trees involved was enriched by the help and technical guidance of Dr. Wayne Clatterbuck, assistant professor of Forestry, Wildlife, and Fisheries with the University of Tennessee Agricultural Extension Service. His comments helped to make our writing on the fall colors accurate and understandable.

Special thanks go to Steve Cotham and the McClung Historical Collection of the Knoxville Public Library System and McClung staff members for their help with the historical accounts and research needed for our manuscript. In addition, our deepest gratitude to Cherel Henderson and the East Tennessee Historical Society for support, encouragement, and untiring assistance in supplying historical information about our wonderful region of East Tennessee.

We must also warmly acknowledge the help of Dr. Don Byerly for his suggestions and technical advice on the geology of East Tennessee. His stylistic suggestions were of great benefit to our text. In addition, we acknowledge the work of past and present geologists in East Tennessee who have deciphered the mysteries of the complex geologic history of the region.

Also, we appreciate the *Knoxville News-Sentinel* for allowing us to use some portions of material written by Fred Brown that was previously published by the *News-Sentinel.*

Our sincere and special thanks go to Jeanne McDonald, our editor, whose advice, tireless effort expended on this manuscript, and suggestions guided and transformed this work into its final form. We also want to thank Jennifer Siler and the staff of the University of Tennessee Press for their patience in seeing this project through to completion.

As always, we are grateful to our wives, Alice Ann and Jeanne, for their patience and understanding, especially because of our absence on the weekends and holidays that were consumed by field trips, research, and writing.

Introduction

Fall in East Tennessee is too beautiful to spend indoors. It's a time to explore, to enjoy the changing landscape, a time to view the region's natural treasures from an entirely new perspective. The fall colors, an annual display of varying intensity, show themselves every October for three or four weeks, but the mountains are always changing, although generally in ways so slow that the span of our lifetimes would have to increase dramatically for the human eye to notice.

For the artist or the Sunday driver, the canvas of autumn leaves is spectacular. For the geologist or amateur scientist, the mountains themselves are the favored attraction. Tossed like stones through millennia, these peaks are storehouses containing records of the past, tracing the movement of ancient rocks assembling and disassembling, rising up from ages-old seas, and then sinking into deep plains.

For example, at the state's new scenic overlook in Unicoi County, located about seven miles into Tennessee from the North Carolina state line, it is possible to face in any direction of the vista and see a multitude of mountains rippling in rhythmic profile, their peaks punching through clouds. Hardwoods that cling to the mountainsides wear leafy crowns of sumptuous colors, flaunting their autumn finery against the darker definition of evergreens.

But the view at Unicoi's overlook also reveals the result of a thrust fault, thousands of feet deep, that was created by old rock slamming into young rock, and then climbing over it when the continents of Africa and North America rammed into each other 230 million years ago. In this area, the fault marks the boundary between the Blue Ridge Mountains and the Valley and Ridge section of Tennessee.

It's not possible to see the younger rocks, but the older rocks are still visible mountains that sweep around this panorama. They were formed from ocean sediments thought to have originated some two hundred miles southeast of East Tennessee and displaced landward with the closing of the proto-Atlantic Ocean (Hatcher, 1989). These ocean sediments, which later became layers of rock, came from somewhere around Charlotte, North Carolina, on the bottom of an ancient sea floor where tremendous sharks once swam. The mountains were pushed eastward in a remarkable event.

Even more startling is the fact that as the continents began separating, about the time of the earliest dinosaurs, the Appalachians were already towering mountains containing some of the oldest rocks in the eastern half of the United States. They are not the oldest—they would have to be close to three billion years to claim that distinction—but some of the rocks of East Tennessee are in the one-billion-year-old range, giving them earthly bragging rights as hoary sentinels of time.

Everywhere travelers go on these fall color trips, they will find a wealth of color, but by paying close attention to the topology, it is possible to learn more than geology. The topology reveals how the land shaped the very history of the East Tennessee area.

In fact, East Tennessee was shaped by its geologic history long before the arrival of human beings. And when humans did come along, they had to learn to live in concert with the land. Many of the legends of the early Indian tribes deal with the land, sun, water, and plant and animal life. Nature was a force to be respected, and settlers had to adapt or die off. In the end, Native Americans were unable to adapt to a new kind of animal arriving on the East Tennessee landscape: the mainly Scotch-Irish pioneers, the long hunters, who were bravely exploring worlds that were new to them. They followed ridgelines to the mountain gaps, which wound down into the river valleys, where they cut down ancient trees to establish log homes close to sources of water and game. Some early pioneers believed that living on the ridges was better for the health: the air seemed fresher and cooler, the sun was more abundant, and ridges were closer to heaven. The human history of East Tennessee is comparable to the geologic history of the mountains, the slow struggle to carve out territory and the constant push for supremacy. There is a wealth of prehistory and history in the area, more than enough to satisfy the history buff, amateur geologist, botanist, biologist, or tourist.

But East Tennessee was shaped more by the marathon of its mountains than by the footrace of human habitation. Before people walked here, time had left its imprint. Later, when human beings did put their stamp on the mountains, it was with explosives and huge machinery. Mountains cannot be moved, but they can be carved. Before that, nature did all the work as pulsing rock strata pushed toward the clouds, and erosion dropped valleys in place between the ripples. But mountains that have been sliced away to make highways can actually help geologists read the physical nature of their history and help the tourist to examine the past life of the rocks.

In East Tennessee today, visitors can walk along what were once ancient shorelines, lagoons, and barrier islands. They can see dwarfed ferns that once stood as tall as trees and toss cobblestones that were once as big as mountains. They can dip their fingers into a stream eddy and find gravel that once clung to the floor of a bog from which now-extinct animals drank.

The ultimate road trip for any close observer is to find his or her own place in time, to understand the conjunction of natural and human-made history of a particular mountain or road and understand how each affected and influenced the other in the creation of East Tennessee.

In a few years—say, a million or so—this landscape will be gone. We are eroding the hills, taking the down elevator, and now we are in the ancient valleys.

In the meantime, standing on top of one of these boundaries, it is possible to stare into both the past and the present and glimpse this remarkable union of human being and nature. And in the fall, particularly October, these mountains will all be dressed in their most colorful attire, waiting for the celebration of their enduring beauty.

By seeing fall color—that spectacularly beautiful phenomenon that lasts for a mere couple of weeks—in its proper historical and geological context, paradoxically, the careful observer can get perhaps as close as possible to time without end in East Tennessee.

Color

Although the tree leaves in southern deciduous forests burst into brilliant color every autumn, the beginning of that remarkable change begins earlier—on June 21, the longest day of the year and the start of the summer solstice.

It's partly true that tree leaves alter their colors because of chemical changes. It is more correct to think of the change in terms of combinations of chemicals and conditions. The ingredients that triggers the change is the length of day, in alliance with elevation and temperature, which are two most important components. The greatest influences on color are shorter days with primarily cooler temperatures and less moisture, resulting in changes in leaf pigment.

With the onset of the solstice, the earth's tilt has its maximum effect on the amount of daylight we in the Northern Hemisphere experience. As the earth continues its orbit around the sun, the daylight hours start decreasing, until the fall equinox (around September 22–23), when we get equal amounts of daylight and dark. By December 21–22, the earth's tilt reaches its maximum point away from the sun (the winter solstice), resulting in the shortest daylight hours. The 23.5-degree tilt of the earth as it orbits the sun causes the annual changes in light and temperature and the four seasons.

During the fall equinox, with its shorter days and less-direct sunlight, weather is cooler, and the change in temperature begins to speed up the internal changes in the plants that soon begin to produce the brilliant leaf colors in our area, around the third week of October.

As for the chemistry of color changes, a sensory mechanism in tree cells begins when change in the length of daylight begins. The major color element in plants is chlorophyll, the greening ingredient. Plants absorb carbon dioxide and transform it into sugars that feed the plant. The warmer the temperature is, the more sugar the plant makes; and the more sugar produced, the more energy is absorbed by chlorophyll in the photosynthesis process that produces food for the plant's growth.

When cool nights arrive, the amount of chlorophyll drops off, and the leaf begins to age as the plant's stem begins to squeeze life out of the leaf, halting the flow of sugar.

In addition, carotene and xanthophyll, pigments that reflect orange and yellow light respectively, are present in the leaf in such small amounts that they are masked by chlorophyll during the growth season. This is one reason why green leaves change to yellow as the weather gets cooler and the amount of chlorophyll decreases. Leaves usually go from green to yellow, though a few trees, including some maples, do not stay in their yellow phase for very long.

In some leaves, when sugars and other compounds begin to break down, they are converted into anthocyanins, one of the elements necessary to make an apple red or a grape purple, as well as red leaves. As chlorophyll in these varieties breaks down, the leaves turn orange, red, maroon, or bright red. Cool nights and bright sunny days produce even more anthocyanin, so the more slowly the leaf dies, the brighter its colors.

Understanding which trees turn what color is easy once the forest is considered as a whole. Tulip poplar, birches, beech, and hickory trees are always yellow, while oak, maple, dogwood, sourwood, sweetgum, and blackgum are almost always red. Some leaves can reach a deeper red, and under certain conditions, the leaves may have a yellow cast.

Sassafras, red maple, and sugar maple put on a dazzling display of color every year. These trees hit high gear when they change from green. Sugar maple leaves are usually brilliant yellow, but some can become orange. Red maples are just that—red. But the maple family can also be red, yellow, brown, or some combination of these colors. These trees are the real show-offs of autumn.

In East Tennessee autumn tends to make its presence felt everywhere. Combine the hardwood trees with an abundance of evergreens—hemlock, fir, spruce, pine, and cedar—and the makings of a gorgeous wildwood palette are everywhere to be seen. Against the contrast of evergreens, reds seem even redder.

In North America fall usually begins in the Longfellow Mountains of Maine and marches south, running through the White Mountains of New Hampshire and then through the Green Mountains of Vermont in about mid-September.

The Adirondacks usually have about the same seasonal schedule. Then fall comes to the Catskills, the Poconos, the Cumberlands, the Alleghenies, the Monogahelas, and finally the Blue Ridge. About mid-October, fall hits the Great Smoky Mountains in full force.

Autumn colors in Tennessee begin first in the higher elevations, responding to their cooler temperatures and to shorter days. Then the fall color simply seems to slide down the mountainsides, progressing to the lower elevations and providing a full canvas of color. The average peak period of fall in Tennessee runs from the third or fourth week in October to the first week in November, but by early November the show is definitely over.

There is no fall like the fall in the Southeast. The leaf colors are so dramatic in the South because it has a greater diversity of hardwood trees than just about any other region on the planet. Traveling the backroads is one of the very best ways to view fall colors because on the edges of the forest the greatest diversity of trees can be found.

The best vistas are in the foothills and high mountains of the Appalachians including the Alleghenies and Cumberlands in the South and following the Alleghenies to the North that form the spine of eastern North America from north Georgia to Quebec. And back roads are the best because they are usually less traveled and have an older border of trees, providing a mature canopy along the roadway, than the interstates.

The roads described in this narrative range in size and complexity from six-lane interstate highways to two-lane state highways and to one-lane, one-way gravel roads. But even the gravel roads are passable in a car. However, bad weather in the colder months may close some of the backroads even to four-wheel-drive vehicles.

Fig. 1.
A typical interstate highway marker.

Fig. 2
U.S. highways are indicated by a white shield with black numbers.

Apart from the interstate highways, most of the routes outlined in this book are two-lane paved roads that will take the visitor into the backcountry of East Tennessee. Most are state highways and tend to be somewhat curvy and hilly. Many do not have pull-off shoulders, so use caution when slowing down to view trees and their colors.

Interstate highways, federal highways, and state roads are usually marked with identifying signs. The interstate system uses a red, white, and blue shield with the interstate numerals across the face of the shield (fig. 1). Every federal (U.S.) highway is marked with a black and white shield, which has the number of the particular highway on it (fig. 2). The federal highways usually run across state lines—for example, U.S. 441, which continues from Tennessee into North Carolina and then into Georgia and Florida, or U.S. 70, which continues into North Carolina from Tennessee.

The Tennessee state highway system includes both primary and secondary roads. A rectangular, black-and-white sign marks a primary highway with the highway number inside the rectangle (fig. 3). Primary highways are usually main arteries connecting towns and cities. In some cases such highways are coincidentally U.S. highways (for example, U.S. 11W is also Tennessee State Route 1). A triangular, black-and-white sign identifies the secondary highways,

Fig. 3.
The primary highways in Tennessee are marked with a white rectangle sign containing black numerals and the word Tennessee. Some of the more scenic Tennessee routes are indicated with the Tennessee scenic parkway sign containing the image of the state bird, the mockingbird.

with the highway number inside the triangle (fig. 4). These roads normally carry a smaller volume of traffic and are usually found in rural areas where curvy roads are more common (figs. 5 and 6). The secondary highways are connected to the primary highways.

County and urban roads are not generally identified except by local names on road signs, and unfortunately some of these roads do not have names or, when they do, may not always be marked by visible signs.

The landscape across East Tennessee is varied, challenging, and beautiful enough to have enticed early settlers to explore the region. Geomorphologists (geologists who study the earth's surface features) have classified the landscape in East Tennessee into three subdivisions: Blue Ridge, Valley and Ridge, and Cumberland Plateau (Klepser, 1967: 73–77; R. L. Wilson, 1981:5) (fig. 7). These physiographic divisions usually coincide with major changes in topography and geology, and each of these divisions provides unique opportunities for viewing fall colors.

Fig. 4.
Tennessee secondary highways are indicated by a white triangular sign containing black numerals.

The Blue Ridge consists of the high mountainous regions of extreme eastern Tennessee, including the Great Smoky Mountains National Park and the Cherokee National Forest. Precambrian-age metamorphosed sedimentary rocks, as well as some igneous rocks, ordinarily underlie the terrain (table 1). In a few places in the Blue Ridge, Paleozoic

Fig. 5.
Most of the roads in the rural areas are two lane but offer excellent fall color viewing.

Fig. 6.
Some of the rural and mountain roads are very curvy and cautious driving is very important.

Fig. 7.
This schematic and very generalized block diagram illustrates the complex structural arrangement of the rocks found in the Blue Ridge, Valley and Ridge, and Cumberland Plateau provinces.

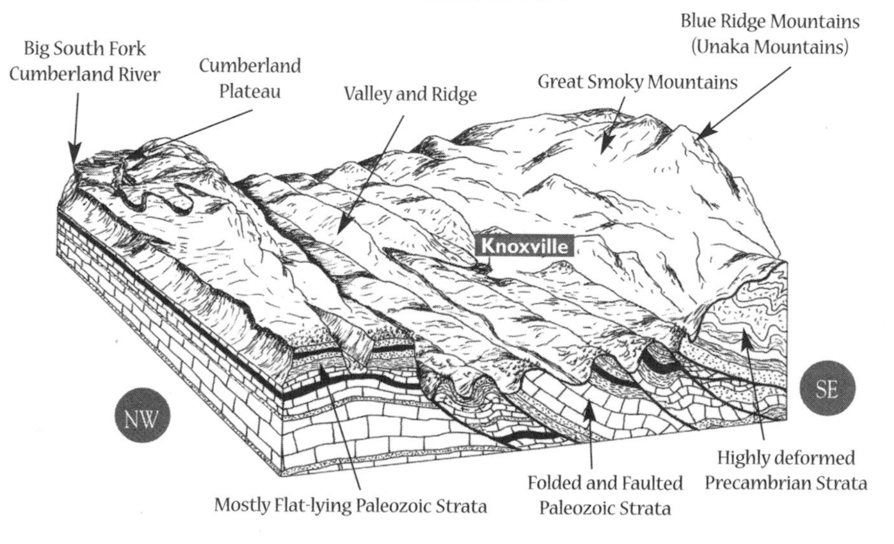

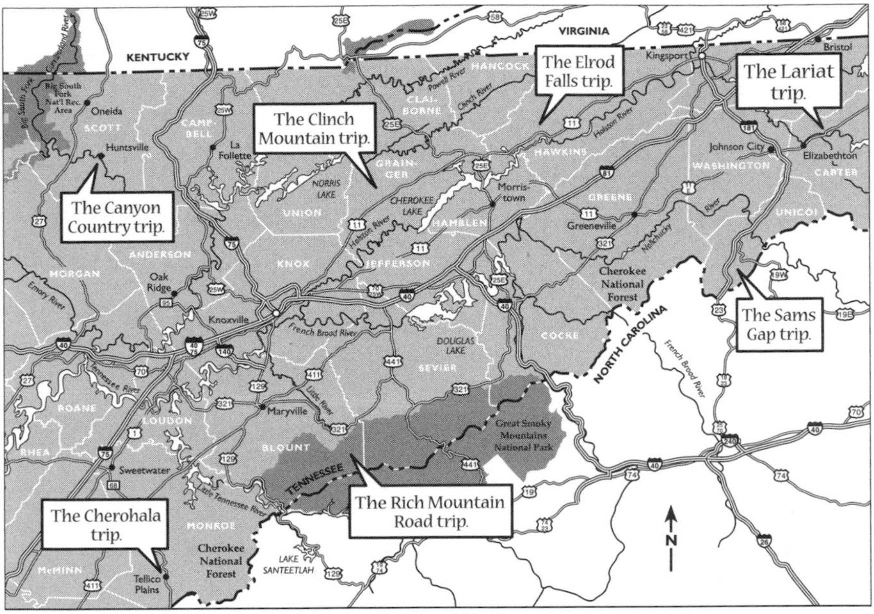

Map 1.
Tennessee and the trip locations in East Tennessee.

rocks are exposed, primarily along the western front of the Blue Ridge and in isolated coves.

The Valley and Ridge is found primarily between the Cumberland Plateau to the west and the Blue Ridge to the east. It incorporates parallel ridges and valleys underlain by folded and faulted sedimentary Paleozoic-age rocks. The major metropolitan areas of East Tennessee are located in the Valley and Ridge and include Knoxville, Chattanooga, Johnson City, Kingsport, Greeneville, Sweetwater, Cleveland, and Morristown.

The Cumberland Plateau, located to the west of the Valley and Ridge, is commonly a high, tabletop-like landscape bounded on the west and east by escarpments 1,000 feet or more in height. The plateau varies in elevation from about 1,000 to 3,000 feet and is underlain by mostly flat-lying Paleozoic-age sedimentary rocks. Coal is often present in the plateau region.

The tours described in this work begin with trips into the Blue Ridge high country. These four trips are followed by two trips in the Valley and Ridge topography. The final excursion takes us to Canyon Country, where the plateau environment literally jumps out in color during the fall. Most of the trip descriptions use downtown Knoxville as a starting point (map 1), with the exception of the Lariat, Sam's Gap, and Rich Mountain Road. These three trips begin at points outside of Knox County.

Time Units of the Geological Time Scale					Development of Plants and Animals
Eon	Era	Period	Epoch		
Phanerozoic	Cenozoic	Quaternary	Holocene	0.01	Humans develop
			Pleistocene	2	
		Tertiary	Pliocene	5	"Age of Mammals"
			Miocene	24	
			Oligocene	36	
			Eocene	58	
			Paleocene	65	Extinction of dinosaurs and many other species
	Mesozoic	Cretaceous		140	First flowering plants
		Jurassic		205	First birds Dinosaurs dominant
		Triassic		240	Extinction of trilobites and many other marine animals
	Paleozoic	Permian		290	
		Pennsylvanian		330	Large coal swamps First reptiles
		Mississippian		360	Amphibians abundant
		Devonian		410	First insect fossils Fishes dominant
		Silurian		435	First land plants
		Ordovician		500	First fishes
		Cambrian		570	Trilobites dominant First organisms with shells
Proterozoic		Collectively called Precambrian, comprises over 85 percent of the geologic time scale			First multicelled organisms First one-celled organisms
Archean					Age of the oldest rocks
Hadean		4600			Origin of earth

Table 1.

Generalized Geologic Time Scale. Adapted from the American Geological Institute (age represented in millions of years).

Driving across these varied landscapes during the fall season provides a glorious and rewarding color experience. The plateau, valleys and ridges, and high country are all vegetated with a mixture of evergreen and deciduous trees that paint the landscape with rich hues of green, red, yellow, orange, and gold during the month of October. Traveling these back roads of East Tennessee allows the visitor to explore not only the colors, but also the natural and cultural history found along each route.

To make the trip more informative, it's important to understand which leaves go with a particular tree. The short guide below displays a leaf and gives the name and Latin designation of the corresponding tree. With the aid of a reliable tree-identification book, no one should have any trouble picking out the most colorful leaves and trees.

Common Trees and Their Colors

Following is a list of some of the more common trees found in East Tennessee that display fall colors. (Lindsey and Bell, 1983; Little, 1979; Radford, Ahels, and Bell, 1983).

Beech

Fagus grandifolia

Found throughout the Southeast, this tree turns a rich yellow to bronze. The leaf, which is elliptic with long, evenly spaced veins, usually stays on the tree well into winter, but eventually turns brown. When the autumn sun hits this tree, it bursts into a rich yellow with just a pinch of orange (fig. 8).

Fig. 8. *The beech tree is common in the hardwood forests of East Tennessee. The sharp, serrated leaf edge is characteristic, as well as the yellow to orange rust color.*

Ironwood

Carpinus caroliniana

Prevalent in southern deciduous forests, its small but distinctive double serrated leaf turns from red to orange. Sometimes called "musclewood" because of the muscle-like ridges under its smooth gray bark, the tree is found along streams and usually grows below three thousand feet in elevation.

Tulip Poplar

Liriodendron tulipifera

Tennessee's state tree, the tulip poplar has one of the most distinctive leaves of any tree in the southern Appalachians—wide and deeply notched. It changes from green to light golden yellow in the fall (fig. 9). The symmetrical, cone-shaped poplar filled in when the chestnuts died in a blight more than fifty years ago, and the trees now speckle the hills in broad, even stands. Their distinctive yellow helps to emphasize other colors, such as reds, browns, and greens. Poplars are found in moist sites, and the ancient "pure stands" of poplar are found on old farms or in abandoned pastureland that is at least fifty years old.

Shagbark Hickory

Carya ovata

The golden yellow of hickories produces one of the most dazzling shows in the forest. The yellows actually seem to vibrate when wind rustles through the trees. The shagbark gets its name from its shaggy strips of light gray bark. Four of its five

Fig. 9.
The state tree, the tulip poplar has a characteristic shape and yellow leaf.

leaves are elliptical. The shagbark's cousin, Pignut Hickory (Carya glabra) is found widely throughout southern Appalachia. Other hickory trees found in Appalachian woodlands are Mockernut Hickory (*Carya tomentosa*) and Bitternut Hickory (*Carya cordiformis*), which are yellow.

White Ash

Fraxinus americana

White ash is one of the more spectacular showoffs of autumn. Its leaves, seven to the stem, run the gamut of colors from yellow to red to maroon to brown. These leaves seem to burst out of the color chute first, while other hardwoods have yet to produce any colorful leaves. This tree is found in most moist woodlands. Its wood was once used for buggy wheels, baseball bats, and skis.

Sassafras

Sassafras albidum

Sassafras is one of the greatest performers on autumn's stage. Found along fencerows, roadsides, and the forest margins, the sassafras has a variety of leaf shapes. Some are lobed on one side and others look like a mitten, with one lobe or no lobe at all (fig. 10).

Fig. 10. *The sassafras tree leaf may be light red to bright yellow and may contain from one to three leaf lobes.*

Fig. 11.
The red maple has a characteristic multipointed, starlike leaf that is usually bright red.

Sugar Maple

Acer saccharum

The sugar maple is synonymous with fall. In its spectacular show of color, it spans the color spectrum, from yellow to orange to red. Its wide, distinctive leaf can light up an entire mountainside. No other group of trees exhibits the brilliance and beauty of the maple in southern Appalachian forests. It is the princess of pure and brilliant color.

Red Maple

Acer rubrum

The leaves of the red maple are mostly red, but they can also turn a deep orange (fig. 11). Although the Red Maple is smaller than its sister, the Sugar Maple, it makes up in color what it lacks in size. It is found from Newfoundland to southern Florida.

Boxelder

Acer negundo

Another member of the maple family is a tree whose leaves in the fall look as if they have been painted with soft yellow pastels. The leaf of the Boxelder may be bronze, brown, or yellow, or touched with a little bit of all three. Each stem has three to seven leaves. The Boxelder is found in lowland forests and along roadsides.

Sweetgum

Liquidambar styraciflua

Even the most tree-challenged visitor cannot fail to miss this one. The Sweetgum leaf is star-shaped, and in autumn its leaves can run from dark maroon to brilliant red, soft pink, orange, or even clear yellow (fig. 12, opposite page bottom). The tree can reach heights of one hundred feet and is found in low-lying woods or along stream banks at lower elevations, usually below twenty-five hundred feet. In colonial times, the gum of the tree was used as a medicine.

Blackgum

Nyssa sylvatica

The Blackgum tree has one of the brightest red leaves in the forest. Its leaves are long, varying from elliptical to oblong. What makes this tree unique is its penchant for nestling alongside pines, usually in moist soils. Bees love this tree for its nectar.

Buckberry

Gaylussacia spp.

One of the most common sights in the Great Smoky Mountains National Park is the Buckberry, called Bearberry by some old-timers. A shrub in the understory, it covers the ground under oaks and Mountain Laurel at middle-to-high elevations. The colorful Buckberry, whose leaf runs from bright pink to red, is found primarily in southern Appalachia.

Sourwood

Oxydendrum arboreum

More than any other tree in the forest, the Sourwood is symbolic of mountain culture. Its poetic name is the subject of many a southern Appalachian song. By late August, the Sourwood begins its explosion of color, spraying the sides of the Blue Ridge Parkway with its soft reds. Sourwood continues to change into October, when its leaves become a deep, dark red. Sourwood honey is one of the treasures of the mountains.

Fig. 12.
The sweetgum tree has a five-pointed leaf that can range in color from bright yellow and red to deep maroon.

Fig. 13.
A soft pastel red color and bright red berries distinguish the flowering dogwood leaf.

Fig. 14.
The rounded lobes of the white oak tree distinguishes it from the pointed leaf tips of the scarlet oak.

Flowering Dogwood

Cornus florida

The small, lovable dogwood announces spring with its pure white blossom, but it also puts on a fine display in autumn. The tree's fat little berries begin to turn red weeks before its leaves begin to change to red or maroon (fig. 13). It can be found in most southern yards, as well as in woodlands throughout most of the eastern United States. It is a favorite ornamental in the Deep South and is especially revered in East Tennessee.

White Oak

Quercus alba

Oaks represent one of the most varied groups of trees in the Appalachians. Their leaves vary from yellow to dark red and brown. Because the leaf is high in tannin, the leaves of a single tree may go from green to brown, or a single leaf may have a dab of green, a dash of red, or a splotch of brown. The White Oak is usually dull red, but can turn bronze, yellow, or brown. It has one of the most distinctive leafs of all the woodland trees (fig. 14). Acorns from the White Oak are an important food source for the legendary black bear of the Great Smoky Mountains, as well as for squirrels and turkey. They were also used for food by Indians.

Northern Red Oak

Quercus rubra

One of the more familiar trees in the rich mountain soils at lower elevations is the Northern Red Oak. Its leaf, which is triangular in shape, changes from green to red to greenish yellow or brown. On sunny autumn days it actually seems to glow. There are actually two oak families—red and white. The red oaks include northern red, southern red, black, scarlet oak, and others. The white oak family includes white, post, chestnut, and others.

Scarlet Oak

Quercus coccinea

Few trees can match the Scarlet Oak for its fall color. The leaves all turn scarlet almost simultaneously and remain on the tree until they turn brown in late fall. This tree can survive just about anywhere, and when autumn arrives, the angular Scarlet Oak leaf is spectacular.

Virginia Creeper

Parthenocissus quinquefolia

Although the Virginia Creeper is a vine, it contributes brilliantly to the fall color in East Tennessee. The creeper has five leaflets, and its vine rambles along fencerows, stonewalls, and roadsides and can be seen climbing to the upper limbs of trees of the South.

PART I

The Blue Ridge

The Lariat

Destination: *Shady Valley and Backbone Rock.*

Route: *I-181, S.R. 67, U.S. 421, and S.R. 91; begins and ends at I-81/ I-181 interchange (Sullivan, Washington, Carter, and Johnson Counties).*

Cities and Towns: *Johnson City, Elizabethton, Mountain City, and Shady Valley.*

Trip Length: *Approximately 136 miles roundtrip from I-81/I-181 interchange.*

Nature of the Roads: *Four-lane Interstate to two-lane curvy, paved roads; some mountainous sections.*

Special Features: *Mountainous Blue Ridge topography, Watauga Lake, Shady Valley, and Backbone Rock.*

The Lariat tour passes through two physiographic provinces, the Valley and Ridge and the Blue Ridge. The more obvious province in East Tennessee is the Blue Ridge, a vast mountain range that is actually a result of the Alleghanian Mountain–building episode that took place some 230 million years ago. Not only is there magnificent color on this trip, there are also various rock types that will aid in understanding how this part of East Tennessee has been shaped, including sedimentary rocks such as limestone, sandstone, shale, and metamorphic rocks like quartzite, formed long ago during the Cambrian and Ordovician periods. Geologists measure enormous amounts of time the way most of us look at our watches. Instead of seconds, minutes, days, weeks or months, they deal in thousands and millions of years. To unravel those huge spans of time, they have devised their own ways of studying them.

For geologists, the past is divided into four main units of time: the Precambrian, beginning with the origin of earth until about 570 million years ago, a span of nearly four-fifths of all geologic time. Next is the Paleozoic Era, which has seven subdivisions, two of which are represented on this trip: the Cambrian, ranging in age from 500 million to 570 million years ago, and the Ordovician, from 435 to 500 million years ago (see table 1). Most of the rocks in East Tennessee are Paleozoic in origin.

After the Paleozoic Era came the Mesozoic Era, a time span of some 175 million years, which marked the beginning of several new species and groups of animals and the demise of others. Most famous among the animals to disappear during the Mesozoic Era were the dinosaurs.

The Cenozoic Era is the last and most recent chunk of geologic time that geologists have established by studying the earth. This era represents about 65 million years up to the present and features the rapid development of mammals and that most special mammal—the human being.

In the Valley and Ridge Province, so called because of the alternating Valley and Ridge topography, rocks are mostly Cambrian and Ordovician in origin and consist mainly of sandstone, limestone, and shale. The portion of the Lariat tour that enters the Blue Ridge crosses rock strata of the early Cambrian period, which contains quartzite, shale, and sandstone, with occasional exposures of limestone and dolostone, a rock made up of the mineral dolomite.

It was in these mountains of upper East Tennessee where the first Scotch-Irish pioneers immigrated in the mid- to late 1700s. They came from Pennsylvania, down through the Shenandoah, into North Carolina and the frontier. The British, happy in most cases to see pioneers leave occupied territories, called this hardy bunch "the yelling devils." They ranged freely, exhibiting the same fierceness and independence they had utilized in fighting the British during the Revolutionary War, and finally settled along the hills and valleys of the Watauga River Valley. There they formed the first independent government ever established on the American continent.

The Wataugans named one of the first counties of East Tennessee for Landon Carter, architect of one of America's early experiments in self-determination. Carter built a home in Elizabethton in 1772 that was most likely the very first built in what is now the state of Tennessee.

The rich beauty of the river valley was also a brief settling place for Abraham Lincoln's father, and here Daniel Boone is reputed to have killed his first bear. The region's hills and valleys have been sculpted by the languid wanderings of the Watauga River, which shaped the fertile bottomland into a quilt of fruitful fields.

Siam Valley, the outline of which follows the graceful meandering of the river, leads to one of the most enchanting places in the area—the Horseshoe Bend of the Watauga River.

Some of the earliest settlers, known as the People of the Horseshoe, were among the very first white people to migrate to this county from North

Carolina, carrying their belongings on their backs. Their women and children followed with the dogs and cattle. Fearless but cautious, courageous but not foolhardy, the People of the Horseshoe were serious settlers, not vagabonds. They entered just ahead of the first great wave of westward migration and settled on land surrounded by the curve of the river's huge horseshoe bend.

The area's single-wagon road evolved from an Indian trail that had been followed for hunting game. Next it was used as access to the island of Horseshoe Bend, where a community became established. Its people, predominantly Scotch-Irish, isolated themselves from the mainstream of development.

The Watauga Valley is a geologist's paradise. Here the varied rock formations have determined the shape of the land itself, as well as serving as the grand architects of the region, determining just where the first settlers could find sustenance and shelter.

Here it is easy to see the origin of the rock strata and the geologic history. Rocks are tilted and folded, and in some instances faulted. The drive along the route features not only vibrant trees, but also impressive rock formations. Rock layers will appear upturned in most places, with dip angles varying up to 90 degrees from horizontal. One way of thinking about a dip angle is to imagine a stack of boards lying horizontally in a garage. Pick up one end and tilt the stack 45 degrees. The flat surface of the boards represent the rock beds or the dip surface, and the angle from the end of the stack to the floor represents the dip angle. A large amount of the rock in East Tennessee has a dip angle of from 20 to 75 degrees. In some places in the valley, however, the rocks are both vertical and horizontal, which is an unusual feature.

For the early pioneer, these geologic characteristics acted both as barriers and as sources of shelter, water, and wood for fuel and log homes. With these amenities at hand, a settler needed only to work hard and be a good shot to thrive, because there was also much game.

Remnants of life even older than the mountains themselves are present in occasional outcroppings where fossils of ancient organisms can be found—species of worm burrows, brachiopods, and graptolites. Like X-rays of bones and inner organs, these organisms are the final shadows of animals that once lived in an ocean environment and are now frozen in time. They can still be seen in shale, limestone, and sandstone rock of the region, but it takes a great deal of searching.

Traces of one species of animal, *scolithus*, thought to be an actual worm burrow, is sometimes found in the sandstone strata of the Cambrian-age Chilhowee Group. This animal lived in a Cambrian-age shallow-water ocean bottom about 550 million years ago. At one time, the nonsegmented worms may have been up to eight inches long.

Calcareous shale strata, known to Tennessee geologists as the Sevier Formation (Ordovician age, approximately 450–470 million years old) erodes to a thin, yellowish gray soil just thick enough to support a root mat. In the yellow shale partings are the remains of past life forms that died out many millions of years ago. One group of these animals is referred to as graptolites. These animals were thought to have been "floaters" in the ancient ocean waters and were possible forerunners of vertebrate animals.

As for fall foliage, the rolling hills along Interstate 81 are clad in junipers, often mistakenly referred to as red cedar. Everywhere along I-81 it seems as though the Department of Transportation has planted cedar trees. These trees, with their perfect Christmas tree forms, also reflect regional geology because the appearance of cedar trees (junipers) means that the soil is usually thin—no more than a couple of inches deep—and the bedrock is usually a shale or shaly limestone.

To start this trip, a map of East Tennessee is required. It may be a good idea to bring along a picnic lunch, but sampling mountain cooking, which is ample and plentiful along this trip, is also possible. One restaurant in particular, the Ridgewood, is an old, family-operated barbecue place on highway 37. Although the restaurant is located roughly seven miles off the route, the real, old-fashioned barbecue is worth the trip.

A camera and binoculars would be particularly useful here. Many, many years ago, a basket for collecting wild cranberries would have been equally useful, because there was once a cranberry bog in this region. Shady Valley—yes, a valley—is 600 feet higher than Clinch Mountain (which stretches through portions of Grainger, Hancock, and Hawkins Counties in East Tennessee and varies from 2,000 to 2,500 feet in elevation). But now the bog and its tundra are nearly a thing of the past, and only small plots of cranberries are found today, not plentiful enough to pick and fill a basket.

This route eventually winds up in a high plain meadow, the only one in the state, and many consider this one of Tennessee's most panoramic views. Just down the road, after the breathtaking vista, is the shortest tunnel in the nation.

The route of the Lariat can be seen on map 2. It begins at I-181 from I-81 to Johnson City and then follows highway 67 to Elizabethton, the town where the Wataugans first set up residence.

Here is where the loop of the Lariat begins. From Elizabethton, follow highway 67 to Mountain City, over to Shady Valley and highway 91 back to Elizabethton. This is one of the more memorable fall trips in East Tennessee, traveling some 130 miles from the junction of I-81 and I-181 to Shady Valley and back.

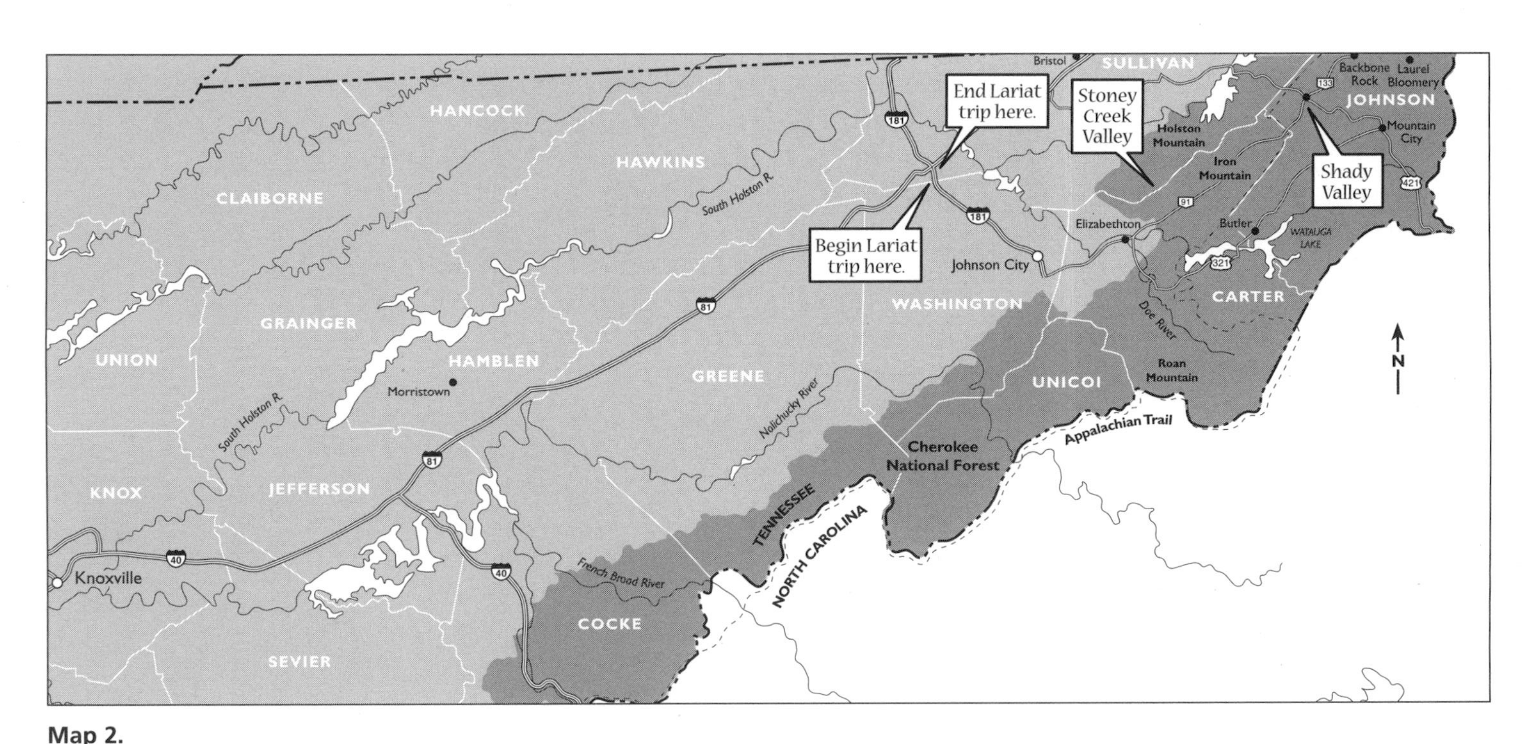

Map 2.
The Lariat trip.

The most interesting part of the trip actually starts as the route enters the "loop," beginning at I-181 toward Johnson City. On the way from Knoxville, some of the fall color can be seen, but the most resplendent part of the trip doesn't begin until the route reaches the "loop" in the Lariat. Still, there are other sights on the way worth seeing. Maples, poplars, and hickories—colors of red, yellow, and gold—are plentiful.

The drive to Johnson City crosses rolling hills and valleys that were formed by the uneven erosion of the underlying sedimentary bedrock. As limestone, sandstone, and shale weather away, the rate of erosion is erratic because the resistance to weathering differs from one type of rock to another. Shale strata erode more easily than limestone and sandstone, and limestone weathers faster than most sandstone, which is the most resistant of the rocks. This explains the uneven landscape in this area.

Along Interstates 81 and 181, valleys and ridges are long and narrow because the sedimentary strata have been folded and ultimately broken by faults as a result of the collision of the continents of Africa and North America some 230 million years ago.

Imagine a stack of "rugs" five deep. Now push one end of this stack toward the other end, so that the mass of rugs moves and eventually hits a wall. The rugs will respond to the resistance of the wall by crumpling into a series of folds that may tend to roll over onto one another, creating a series of elongated parallel folds. The valley and ridges of East Tennessee on this section of the color drive have been pushed and folded in the same manner.

This folding happened to the sediment layers deposited in the seas between Africa and North America and now found in East Tennessee, as well as in Virginia, Pennsylvania, and the northern parts of Alabama and Georgia. The erosion of these "crumpled and folded" sediments has been uneven due to their composition. Sandstone and limestone usually form ridges, while shale strata usually form valleys. That's because of the shale's lesser ability to withstand the erosion processes. Interstate 181 crosses these long valleys and ridges.

From I-81 toward Johnson City on I-181, the road again traverses a series of ridges and valleys. The road cuts across the grain of the tilted bedrock layers instead of lying parallel to the long beds of tilted rock, as it does along I-81. This formation creates the up-and-down nature of the roadway.

In May 2000, State Department of Transportation road crews dug into what could be one of the greatest fossil discoveries in Tennessee history. The spectacular vertebrate fossils were unearthed at Gray near Johnson City along a section of State Route 75. The fossils are from the Miocene Epoch, which began about 18 million years ago and ended about 5 million years ago. The presence of rhinoceros remains in the deposit suggests that the fossils are at least 4.5 to 5 million years old, the general date when rhinoceroses disappeared from North America.

Fig. 15.
The partially uncovered skull of a tapir, believed to be Miocene in age, with the teeth clearly visible. This skull was found during the summer of 2000 at the S.R. 75 road construction project near Gray in Washington County.

Bones of prehistoric tapirs, rhinoceroses, turtles, frogs, a mashed crocodile skull, shellfish, insects, seeds, and tree leaves have been found. In addition bones from a proboscidean (elephant) also were discovered, along with large chunks of tusks. The discovery of entire tapir skeletons and partial skeletons of rhinoceroses excited the scientific world in Tennessee and opened the door for years of study at the new-found site.

Fig. 16.
The remains of a fossil tapir backbone and ribs are shown here from the Gray fossil site. Note the hand lens for scale.

At fifteen miles from the beginning of the trip, leave the interstate onto exit 31, and take highway 67 to Elizabethton, the first side trip of the tour. Not long after entering Elizabethton, Sycamore Shoals State Park can be seen on the left side of the highway. The park is situated where the mighty Watauga River, now dammed and blocked above Elizabethton, once flowed. This is a chance to see the setting for some of the state's most important early history. The state park exhibit relates the dramatic details of the Overmountain Men, who fought and won the Battle of Kings Mountain. During the Revolutionary

War, when the American patriots were all but beaten and the future nation hung in the balance, the Overmountain Men left this very spot to head across the mountains into South Carolina. There they met British troops at Kings Mountain in October 1780 and defeated them in little more than an hour.

The state park is also a good place to get out and walk and, if it's a warm day, to wade in the nearby shoal that gives the park its name. A shoal is a place in a stream where the channel is usually wide and shallow. Shoals can be caused by a number of reasons, but usually they are due to a change in the substrate material and/or the topography that causes the stream flow to slow down. As a result, the water will "drop" any sediment (from silt and sand to cobbles and boulders, depending on the environment) into the stream. Over time, the sediment builds up and creates a natural crossing place in the stream. When settlers crossed over the mountains, this shoal provided one of the first places they could easily ford the Watauga River. A settlement soon developed that eventually grew into Elizabethton.

To quickly rejoin the Lariat route, from Sycamore Shoals, take the highway 67 Bypass around Elizabethton. The town, however, is well worth a side trip, particularly for its architecture. The downtown area still has many of its old brick buildings, some of which have fronts that date to the early 1900s. There is also a famous covered bridge in Elizabethton, on Riverside Road. At the bridge that spans Doe River, fall's splashy reflection is mirrored in the river, with the bridge as a dramatic backdrop. This gorgeous sight itself is well worth the trip, both for the natural beauty and for a look at the old bridge, which was built in 1882. One of the few covered bridges left in the state, it is the only one of this size and stature. This magnificent structure has proved its mettle many times, often withstanding frequent flooding of the Doe River.

Once over the bridge, turn left onto Main Street to return to highway 67. Then turn right and pass through Valley Forge and the Doe River Gorge. Valley Forge got its name from the old iron forge built in the community just where the Doe River emerges from the gorge. Iron ores were extracted from the soft, plastic, or "sticky," clays in overlying residual soil of the Shady Formation, a dolostone rock. The gorge is where the Doe River cuts across upturned and resistent strata of the Chilhowee Group of rock formations. The cut through this rock represents many millions of years of unrelenting flow of the river.

The gorge also marks the boundary between the Valley and Ridge Province and the Blue Ridge Province. There is no mistaking the presence of the mountains now. Steep topography and abundant plant species abound. In a colorful fall season, when all the complicated elements that make up fall brilliance come together, this gorge is wild with color as beech, sourwood, blackgum, and red oak are prevalent.

A few miles on the other side of Elizabethton is a little junction known as Hampton. At this point the trip has gone thirty miles and perhaps forty-five minutes from its beginning. Here, the route crosses the Appalachian Trail. The sign for the trail, not readily seen, is marked by small patches of white paint, usually painted on tree trunks, highway guardrails, and rocks. Hikers get a better view of the markers than fast-moving automobile passengers do, so go slowly to spot them.

This part of the trip is the section where quite beautiful scenery begins as the route continues toward Mountain City. Highway 67 now makes its way into higher elevations. There could even be a chill in the air, which adds a certain briskness to the staggering natural sights.

From Hampton, Watauga Lake will be on the left. This trip should be savored slowly, and stop along the way at one of the roadside pull-offs. The lake area will be ablaze with color, highlighted by the rich greens of hemlock and pine in the background. The scene is like a giant canvas, painted with brilliant hues. The reds and yellows of maple and poplar dominate.

Spectacular color continues for the first eight miles or so as the route winds along the lake. This roadbed once lay under a great river. The road passes by overlooks and picnic areas at the lake, including the scenic overlook at about mile thirty-nine. For a special view, and a spectacular picnic site, wait for Backbone Rock. Across the lake, outcroppings of the hard quartzite bedrock that holds up Iron Mountain can be seen. Here, the Appalachian Trail follows the crest of the mountain, a narrow topped ridge with views down either side into beautiful valleys.

Rock hounds will enjoy trying to spot the cobbles as they leave Watauga Lake. The cobbles are embedded in the banks of the curvy road. They were once buried in the ancient river that covered this area. Geologists call them a terrace deposit because they came from sediment left by an age-old stream, or perhaps by the ancient Watauga River, with the cobbles marking the waterway's course in the earth's history. Many thousands of years have passed, and the stream—or river—is now long gone, but the cobbles of sandstone and quartzite, a metamorphic rock, remain. Metamorphic rocks have been changed dramatically through varying degrees of heat and pressure, creating new forms. For example, sandstone, when subjected to great heat and pressure, will become a metamorphic rock called quartzite. The cobbles seen along the road were most likely carried to this spot from places farther east where quartzite is readily found in the bedrock.

In this area, the bedrock is composed of variegated shale, siltstone, and limestone of the Rome Formation, so-named because the bedrock was first described in Rome, Georgia, and more massive beds of dolostone strata known as the Shady Formation. These strata erode more easily than the surrounding quartzite strata of the Chilhowee Group and tend to form the lower ridges and valleys of the area. The red, maroon, yellow, and pale green rock colors identify the shale, siltstone, and sandstone of the Rome Formation, and most of the limestone and dolostone tend to be a gray color. Because the rock is so soluble, several caves have developed in the rock layers. One of the best-known caves in the area is Dewey Cave, about three miles southwest of Mountain City.

About forty miles from I-81 is one of the extraordinary places in East Tennessee—the town of Butler. Although the original town is now nothing but a memory, crossing the Butler Memorial Bridge is an occasion to recall the former town that is now under tons of water, flooded by the Tennessee Valley Authority in 1948. The town used to exist to the right as you cross the bridge. Early maps show the town of Carderview in the location where Butler is currently located.

Legend has it that Butler's residents were practically washed out of their houses by the Tennessee Valley Authority. The families, though forewarned, refused to leave their homes until the very last minute and departed just ahead of the rising waters, some leaving dirty dishes in the sink and food cooking on the stove. Each year, former Butler residents and their descendants return for a reunion to recall the lives they had led before their town sank beneath Watauga Lake.

As the elevation continues to climb, the colors will also climb in intensity. The route passes through the small, unique settlements of Pine Orchard, Doeville, Pandora, and Doe Valley. The state's Northeast Correctional Center will be to the right just before the road enters Mountain City. At Doeville, note the mountains to the right, which run into Mountain City. At this time of year, Doe Mountain shimmers with seasonal colors, offering one of the tour's grandest backdrops of color.

To continue the route, at Mountain City turn north onto U.S. Highway 421. Going straight on highway 67 will lead to Mountain City, which is a fine side trip. The town boasts many lovely old homes, some with stately columns and sprawling front porches. Farther up the road is Iron Mountain Pottery, where pottery of excellent workmanship, taken right from Iron Mountain, is made and sold.

From Mountain City, the road enters a geologic bowl of color as highway 421 heads toward Shady Valley. Hardwood trees of every description arch overhead as the route winds along the curvy switchbacks. The canopy goes all the way up Iron Mountain, where the road is twisting and narrow, but there will not be a great deal of traffic here: the lucky traveler may be almost alone in a wonderland of fall's splendor.

After crossing over Iron Mountain, the same mountain that was on the left coming into Mountain City, the road heads toward Shady Valley, one of the prettiest valleys in the state. The rocks that support Iron Mountain are of the same rock variety found in the Doe River Gorge. This same formation runs northeast from Hampton and on past Mountain City, so the route now recrosses that highly resistant strike belt, a layer of rock heading off in one direction, and the switchback curves that characterized the way up the mountain will now descend the mountain beneath the same ceiling of color.

About sixty-eight miles from the beginning of the trip, the route enters Shady Valley, a tundra valley that was once quite famous for its cranberries. Today, the only memories of the cranberries are evoked by the valley's annual cranberry festival, which provides revenue to help maintain the Shady Valley School, built from native mountain stone.

The boreal cranberry bogs of Shady Valley, twenty-eight hundred feet above sea level in a bowl surrounded by the Holston, Iron, and Cross Mountains, were once the center of the commercial cranberry industry in the United States, but now this once great bog of ten thousand acres has shrunken to less than one-third of an acre. Jenkins Bog is under the protective custody of East Tennessee State University, since it is the sole remnant of Shady Valley's once superb cranberry past.

Here is what scientists think happened to create the cranberry bogs. When the glaciers ground their way across the planet, covering about 20 percent of the earth's surface, East Tennessee was spared the glaciers, but not the fallout from them. During the last 2.5 million years of glaciation, known as the Ice Age, though there is no evidence the glaciers reached East Tennessee, the extreme temperatures did.

With each major freeze and thaw, sea levels are thought to have changed hundreds of feet. As ice melted and sea levels rose, lower valleys along coastal areas were drowned, increasing soil deposits upstream. Although Shady Valley was not flooded by the rise and fall of the sea, it did experience the climatic changes associated with the glaciers farther north and the sea-level fluctuations. This created a Canadian-type, tundra-like environment of great spans of sphagnum, which makes a watery, acidic nest just right for producing cranberries. The cranberry grows only in treeless areas, mainly where sphagnum moss and lots of water are located. Since the berries do not survive if the overhead canopy shuts off sunlight, they thrive best in sunny glades under very wet conditions.

The cranberries that are left in Shady Valley are not harvested now because they are the progeny of the original cranberries that arrived in Shady Valley in the Ice Age. When the glaciers moved south, they pushed the cranberry south. And when the icy masses retreated north once again, they took some cranberries with them on the return migration.

East Tennessee and Western North Carolina once had quite suitable climates for cranberries. They flourished here, and when the glaciers began retreating back north, some of our cranberries headed north again along with the cooler, wetter climate. As the ice retreated from eastern shores, cranberry seeds began to migrate via birds and wind. Finding new bog areas in the East and South, the cranberries took hold and multiplied.

This is a simple explanation for a complex set of circumstances and occurrences, but it is the one believed most plausible by regional scientists who have studied the cranberry bogs of Shady Valley, where early settlers prospered from the bounty of berries and their wonderful, fertile valley set in the clouds.

Shady Valley's tilt is the result of erosion of a downward fold in the bedrock, known in geologic terms as the Stony Creek Syncline. A syncline is easy to envision. Get a piece of cardboard and bear down on the top so that the cardboard has a section that is bowing upward and another that is bowing downward. The downward bow is called a syncline and the upward bow is an anticline. The Stony Creek Syncline has formed the rocks in an elongated bowl shape. In their structure, the youngest rocks are found in the center (in this case, at the center of Shady Valley), and the older rocks lie farther out from the center. The elongation of the "bowl" structure parallels the structure of Stony Creek Valley and the crests of the Iron and Holston Mountains (fig. 17).

The Cross Mountain Fault cuts across both the Iron and Holston Mountains, near where they seem to converge at the head of Stony Creek Valley, displacing somewhat the rocks in Shady Valley from those in Stony Creek. The rocks in the central part of Shady Valley are layers of dolostone, which tends to dissolve in the presence of acidic groundwater; therefore, the rocks in the central area of the valley have eroded more quickly than those surrounding the valley, causing the bowl-shaped topography of Shady Valley.

For a side trip treat, at the Shady Valley crossroads (fig. 18), where highway 91 and State Road 133 meet, turn right onto highway 133, which leads

Shady Valley

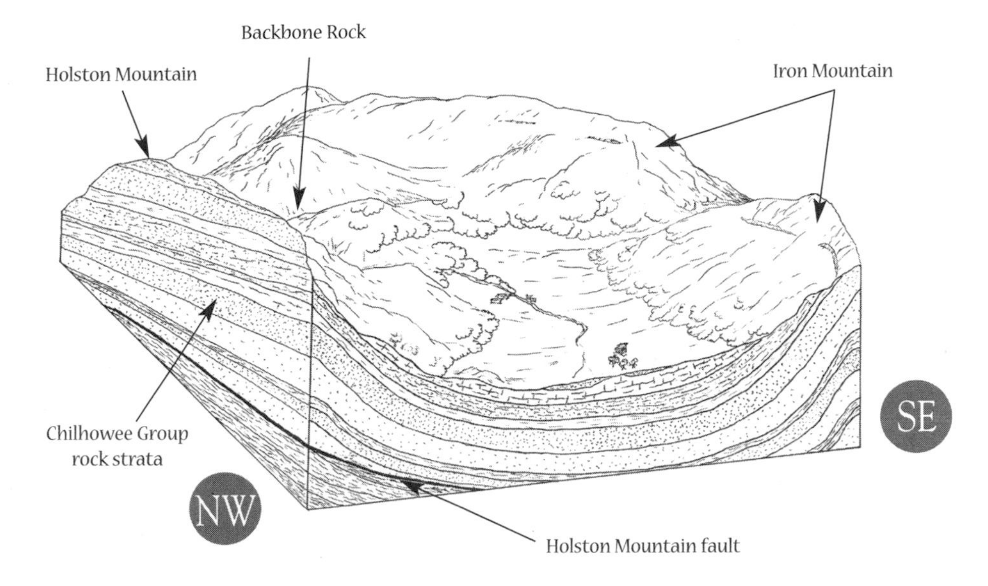

Fig. 17.
Shady Valley in Johnson County is formed in the floor of a geologic structure called the Stony Creek Syncline.

Fig. 18.
Nestled between Iron Mountain and Holston Mountain, Shady Valley has some of the more scenic views of East Tennessee.

to Backbone Rock. At about thirty-three feet is the shortest tunnel in America. The tunnel is carved out of the spur ridge of Holston Mountain and is at the end of this side trip.

For about ten miles out of Shady Valley, the road parallels the small mountain Beaverdam Creek. Here, the trees are so thick they may seem to envelop the roadway, creating a bower of fiery color. This intimate drive leads to the Backbone Rock Recreation Area, the perfect place for a picnic.

The tunnel that runs through Backbone Rock was cut through quartzite around the turn of the century to transport timber and logs to Shady Valley from Abingdon, Virginia. In the 1930s, under President Franklin Delano Roosevelt, the Civilian Conservation Corps built picnic shelters and stone walkways up the mountain, and the U.S. Forest Service refurbished the trails in 1994.

Backbone Rock consists of Cambrian-age quartzite called the Chilhowee Group. This rock, over 450 million years old, is composed almost entirely of quartz. Extremely stable, it weathers very little over time. This narrow ridge of quartzite forms Backbone Rock (fig. 19).

Return to Shady Valley over the same road that goes to Backbone Rock. After leaving Shady Valley on highway 91, heading south, the road passes through Stony Creek, and Holston Mountain is on the right and Iron Mountain on the left. The gorgeous view here is a great finale to the Lariat Loop trip. In autumn, hardwood trees of every description blaze along the hillsides. Note especially the brilliant reds of the blackgum, sweet gum, and sourwood, three varieties that are the reddest in fall. Evergreens will provide a green backdrop to it all, providing breathtaking scenes of fall color at its finest.

Stony Creek gets its name from the numerous quartzite cobbles and boulders found in its channel and banks. The numberless "stones" are remnants of the erosion of the surrounding mountains that are composed of quartzite strata. In future times these same quartzite cobble-strewn deposits may be preserved as a terrace deposit, left high on the side of a ridge overlooking a more deeply incised Stony Creek Valley.

Besides the luminous colors exploding from hillsides and valleys in stunning reds and golds, the gentle rise and fall of the Lariat Loop is punctuated by distant views of rolled hay, fields of goldenrod, and barns of every size and shape. The countryside is dotted with old, white wood-frame farmhouses, roadside stands selling fruit and vegetables, and country stores stocked with ancient mother-of-pearl buttons, licorice sticks, and rock candy.

Fig. 19.
A short but steep walk to the top of Backbone Rock provides the visitor with dramatic views of the fall colors. The top of the rock structure is an exposure of Chilhowee Group quartzite that varies between eight and ten feet wide.

Even the road signs evoke a past that can only be reconstructed in imagination: Big Sandy, Bulldog Hollow, Dugger Branch, Panhandle Road.

Take the time to savor the sights. Ease the pressure on your accelerator. See this area as the settlers saw it—it is ancient, pristine, unspoiled. Here, history seems to blow in the wind and grow from the earth.

Sams Gap

Destination: *Sams Gap at the Tennessee North Carolina state line on U.S. 23 (future I-26).*

Route: *I-40, U.S. 25-70, N.C.213, U.S. 23, Tenn. Route 81, 107, 34, and I-81; begins and ends at I-40/I-81 junction (Cocke, Unicoi, Washington, and Greene Counties, Tennessee, and Madison County, North Carolina).*

Cities and towns: *Newport, Hot Springs, Mars Hill, Erwin, Greeneville.*

Trip length: *Approximately 160 miles.*

Nature of the roads: *Interstate four-lane, two-lane paved highways; numerous curvy sections; one-and-one-half-lane gravel road on forest service road side trip, hilly sections throughout.*

Special features: *Hot Springs, North Carolina, mountainous topography and scenic overlooks in high country.*

The Old English word for the season was "faule," a time of fruition and happiness. Chaucer wrote it first as "autumne," and Shakespeare said the season was "loud as thunder when the clouds in autumne cracke."

Pritchard Metcalf, who lives in a green socket of trees at Sam's Creek on the Tennessee–North Carolina border near Erwin understands that metaphor—loud as thunder in the clouds. He lives in a one-hundred-year-old, "doubleboxed" home: years ago, his kinfolks built the present house around their log cabin. From his home he has watched scores of fogs sweep over the land. His ancestors came here when the forests were thick and rich with wildlife, when trees were so large, the sound they made when they fell to the earth resounded for acres.

Metcalf uses the old ways to predict the severity of fall and winter and how many snowfalls are coming. He relies on a method that is as certain for mountain folks as computers and weather maps for those who live in cities and towns. Although his traditional methods cannot stand up to the analysis provided by modern science, Metcalf is not about to change.

He watches nature. He studies the thickness of the leaves and bark. More leaves mean more fall, he says. It's that simple, and yet that complicated, because it takes the trained eye of someone who has been close to the land.

He counts the number of foggy mornings in September. He watches the thickness of the shucks in the tall, brown corn in his backyard. He looks to the forest floor to observe the carpet of acorns on the ground in the nearby woods. He notes the wooly worms crawling at a dangerously slow pace across the high mountain road in the switchback where he lives. It's going to be a cold winter. He knows that because the wooly worms have already changed from brown fuzz to black.

And when the summer trees are flush with foliage? "Means we are going to have a beautiful fall," Pritchard Metcalf says. Not just pretty. Beautiful.

And the fog? "It's in the fogs and the trees. Trees are loaded with leaves and we've had eight fogs, so that means we will have lots of color and eight snowfalls this winter." Metcalf was right on the money the season he said that. Some years his natural method of predicting the winter misses the mark. But not by much.

Weather computers and scientific research can make very precise predictions these days, of course, but they lack the mountain logic and sense of tradition that comes with Metcalf's predictions.

On this trip, it is possible to see the fogs, the wooly worms, and the corn shucks. And it is absolutely certain that the trees will be in a riot of color, for this is the French Broad River Valley, along the road to Hot Springs and Sams Gap, where Pritchard Metcalf lives.

This route cuts across portions of two physiographic divisions: the Valley and Ridge and the Blue Ridge. By traveling Interstates 40 and 81, U.S. Highways 25, 70, and 23, and several state roads in Tennessee and North Carolina, visitors on this route can have it all: views of beautiful and varied fall vegetation and landscapes with rocks ranging in age from nearly 1.2 billion years old to 400 million years old.

These ridges and valleys, as well as the high mountain landscapes, are propped up by a variety of rocks, which include sedimentary and metamorphic classification. Limestone, sandstone, gneiss, and slate are exposed in great amounts along the route.

Another part of this trip to look forward to is Hot Springs, North Carolina, which has the unique geological distinction of geothermally heated groundwater.

To get started, begin the trip at the intersection of Interstate 40 and Interstate 81. At that point, follow I-40 east to Newport. Go through Newport

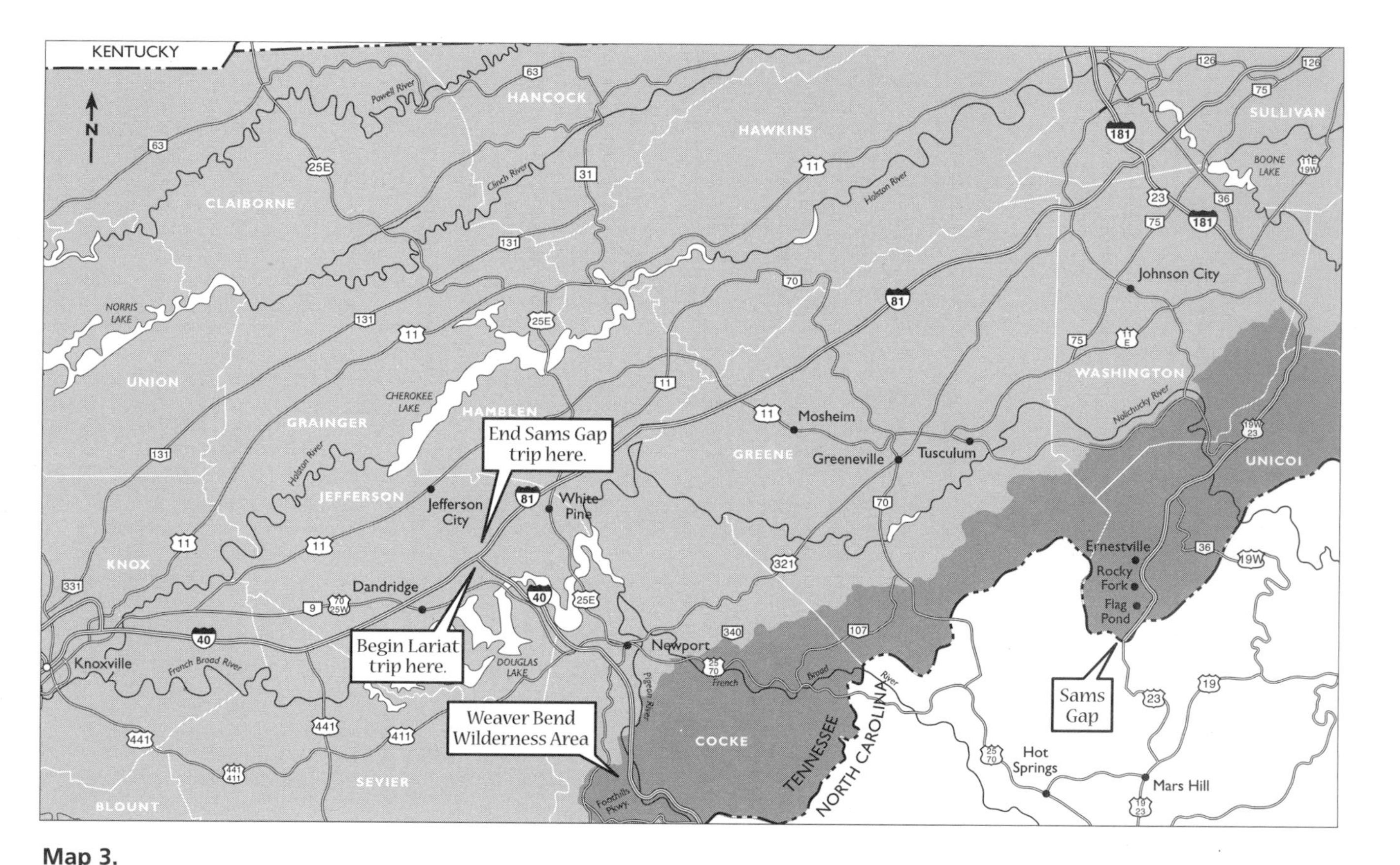

Map 3.
The Sams Gap trip.

on U.S. 70 to Hot Springs (map 3). Stay on U.S. 70 to Marshall, North Carolina, and take N.C. 213 to Mars Hill. At Mars Hill, follow U.S. 23 to the top of the mountain and the North Carolina line. Stay on U.S. 23, newly opened by Tennessee. The views from this new highway, which will eventually become Interstate 26, are nothing less than spectacular.

This trip could take one or two days, either by making it a single-day round trip or by stretching it to two by going on to Mountain City and visiting Shady Valley on the return to Knoxville.

From Knoxville to the junction with Interstate 81, I-40 crosses over the Valley and Ridge topography, which is underlain by Cambrian and Ordovician limestone and shale strata. I-40 typically follows the strike of the strata, trending into long strike valleys and crossing broad ridges underlain by limestone and dolostone mantled with cherty, reddish orange residual soil. In Newport the roadway cuts across the regional northeast–southwest strike of the bedrock. Some three and a half miles from I-81, I-40 crosses the French Broad River (Douglas Lake), where east of the French Broad River the topography is characterized by more rounded knobs and hills. Shale strata, commonly seen along I-40 as light yellowish brown chips of shale, typically underlie these rounded hills.

Outside of Newport on U.S. 70, the autumn show of color really begins to intensify. From the I-40/I-81 juncture, follow I-40 to Newport and exit onto U.S. Highway 70 (exit 432B) about 10.7 miles from the start of the trip. Follow U.S. 70 through Newport and on toward Del Rio, Tennessee.

In Newport the route passes by the Old Millstone Inn, constructed of Lower Cambrian Chilhowee Group sandstone. At the Hunt/Wolf/Van Camp canning plant, the rock bluff that can be seen across the Pigeon River from the plant is composed of the Knox Dolostone, Cambrian/Ordovician-age strata.

About twenty-two miles from the juncture of I-40 and I-81, this trip enters an area covered with colluvium—large blocks of rock that have moved down the mountain slopes, where they form an unstable type of deposit along the road. The deposit is referred to as talus, a collection of fallen material forming a slope at the foot of a steeper bluff or cliff. This unstable material of broken rock debris is also called a block stream.

A serious block-glide type of landslide, also referred to as a planar failure, occurred at this location in the fall of 1995. In this area, the rock strata dip toward the road, where weathering and road construction undercut the strata. The Tennessee Department of Transportation maintenance forces corrected the problem by removing the sliding mass of rock and constructing a rockfall catchment fence.

Between Del Rio and Hot Springs, North Carolina, on U.S. Highway 70, once great, ancient mountains, now eroded to mere remnants of their former magnificence, can be clearly seen. At about twenty miles, an ancient river terrace that helped to shape and cut away those old mountains can be found. The terrace consists of rounded cobbles and boulders suspended in a reddish brown clay-sand.

At twenty-three miles the route intersects with highway 340. Stay on U.S. 25-70 toward North Carolina.

At the Weaver Bend Wilderness Area in the Cherokee National Forest, turn left onto Forest Service Road 209 just before the road follows the French Broad River bridge. The next seven miles has one of the most dazzling fall views to be found in East Tennessee. This narrow gravel road winds through a mixed hardwood forest, where roadsides are dotted with fall asters and blue coreopsis, which bloom until the first frost. Ironweed and black-eyed Susans also abound.

After about two miles, turn right onto another gravel road that eventually leads back to the French Broad River. Cross a railroad track and then turn left to the river. Here is an excellent spot for taking photographs: there is a broad bluff, which bent as it was cut by the river. Across the river the former community known as Weaverville once stood. The town is gone, but in the fall sourwood trees seem to burn brightly in front of the evergreens.

Just above and to the left is an ancient river terrace marking the place where, thousands of years ago, the river ran through—forty to fifty feet above the bluff. The smooth cobblestones here were brought by the river from great distances and left for all time. Think of the cobbles as once great stones, now worn down to pebbles. That is the length of time that is on display here.

This is an excellent site to take a deep breath, feel the wind, and smell the air. The water reflects every color in the trees, just as if they were replicated in watercolor.

Retrace the route back to U.S. 25-70 in Cocke County, one of Tennessee's most economically and socially depressed regions. It is blessed by beauty and damned by its isolation. As a young woman in 1959, Catherine Marshall came here to visit the site where her mother, Leonora Wood, had taught in the 1920s in the deep recesses of the mountains. Her experiences were later described in her 1968 novel, *Christy*, a story that describes the richness, poverty, and the austerity of this area.

The county is shaped like a triangle, with its base nestling up against the Great Smoky Mountains. The land was first settled in 1783 in a lush area known as Irish Bottom. The first people here had names like McNutt,

McNabb, McGuinn, Doherty, and Gilliland (*Goodspeed's History of Tennessee*, 1887). Pennsylvania Germans also found their way here and settled an area that came to be known as Dutch Bottom. Other early arrivals were French Huguenots, Dutch, Irish, Scotch, and Africans, who moved in to displace the Cherokee. Long before any of these landings, however, the Mound Builders had settled on Cocke County's western border, along the river plains. They date back thousands of years.

To the Indians, the rivers were sacred. They named them "Agiqua," which means "broad." The Big Pigeon River, which cuts through the county, was first called "Wayeh," which means "beautiful maiden."

The county draws its name from Lt. Col. Richard Cocke from Devonshire, England. In 1632, he brought twenty families with him to the new country, where he had a grant of some three thousand acres, which included much of the present Cocke County area (*Goodspeed's History of Tennessee*, 1887). Parrottsville, the third-oldest town in Tennessee, was settled in 1769 by John Parrott, the son of Frederick Parrott from Alsace-Lorraine, France. Other colorful community names are sprinkled across Cocke County: Purty Holler Gap, which is near Cosby; Sunset Gap, Irish Cut, Pig Trot, Frog Pond, Bat Harbor, Bizon, Bogart, Boomer, and Broad-Axe Hill. The bluegrass ballad "Crippled Creek" gets its name from Crippled Creek in Hartford (Brown, 1990).

Mount Guyot, one of the highest peaks in the Smokies, rises 6,636 feet on Cocke County's southern boundary, separating the county and the state from Haywood County in North Carolina.

Another place of note in Cocke County is Slabtown, near Del Rio, home of Metropolitan Opera diva Grace Moore. Slabtown is also the residence of the rare and endangered yellow gopher wood, one of the rarest trees in Tennessee and the South. The tree blooms in June, and only three or four have ever been found. In the Book of Genesis, God tells Noah to make his ark of gopher wood, a fact that is not lost on the people of Del Rio, who may suffer from intense isolation, but they are proud of their biblical distinction in the state.

Del Rio, which means "The River," has also produced May Justus, a noted children's book author, and Edmund Burnett, a distinguished scholar.

The county has always had a wonderful mix of people: Anglo-Saxons speaking a distinctive English dialect and people in another area conversing in guttural German-Dutch accents, while Huguenot neighbors who brought in soft French speech patterns.

Although diverse in national origins, religion, and speech, the earliest people here were united in their quest for independence.

Just past Del Rio, the route crosses the French Broad River immediately after passing Forest Service Road 209. Leave Tennessee at about 34 miles as measured from the beginning of the trip. Cross into North Carolina, go through the Sleepy Valley community, and then approach Hot Springs.

Near Hot Springs is a short road with a great view of the Hot Springs area and the French Broad River Valley. This turnoff is at 38.5 miles. Turn left off U.S. 25-70 onto a side road and travel about 100 yards (this road is a section of old highway 25-70 and is about 1,000 feet long). The town of Hot Springs is below the overlook along the floodplain of the French Broad River. When leaving the overlook area, continue around the curve onto the old highway, which will reconnect with the newer highway now in use.

The springs in Hot Springs were discovered in 1778 and have been the center of numerous hotels and spas over the years. Geologically, the origin of the springs remains an open question. The leading theory of how water heats itself involves water entering the subsurface through cracks and fractures in the rock. Some of the fractures are thought to run along a fault known as the Dry Pond Ridge fault, which penetrates downward beneath the surface of the earth for several thousand feet (Byerly, 1997; Oriel, 1951).

The water continues to travel downward through the fractures for up to four to five thousand feet beneath the surface until it reaches the vertical bedded layers of the Shady Formation (rock strata composed of layers of dolostone—a magnesium-calcium carbonate). The geothermal gradient heats the groundwater (at the rate of 30 degrees centigrade per kilometer of depth), and the water eventually reaches a temperature of 104 degrees Fahrenheit (about 40 degrees centigrade) before being pushed back to the surface by hydrostatic pressure along solution channels developed in the vertically bedded strata of the Shady Formation. The heated water surfaces as springs in and along the French Broad River and has been contained for use in spas. In addition to the hot spring water, the area is also known for barite, manganese, and sulfide mineral (zinc, lead, and iron) mines and prospect pits.

Rounding the turns down into Hot Springs brings views of bold splashes of fall color. Hot Springs is one of those quaint, beautiful places that is a reminder of quieter, simpler times. The Hot Springs Spa and Natural Mineral Water can be enjoyed just on the edge of town.

Outside of Hot Springs, pay attention to the broad vistas. This leg of the trip leads to highway 212, then turns right and heads east (this is also U.S. 25-70).

Near mile forty-three, the Appalachian Trail crosses the highway at Tanyard Gap and extends from Springer Mountain in Georgia to Mount Katahdin, Maine—a distance of some two thousand miles.

At forty-five miles, the Millwheel Café appears along the route. It is tucked into the intersection with N.C. 208 and U.S. 25-70. This is a good place to stop: there is good food and a pleasant atmosphere. Then continue following U.S. 25-70 to Marshall.

The experience of fall here is exceptional. Laurel Creek, Shelton Creek, and other springs jump here and there, making their uncommon music, which somehow seems all the more singular during the fall, even though foliage here is not as thick.

Off on a treeline, it is possible to see a row of beehive boxes at the edge of the woods. An occasional field of sorghum flashes by, as well as golden tobacco stalks, slanted like leaning tepees. This is tobacco country, and the farms are dotted with wooden footbridges leading to pole barns brimming with the brown leaf.

At Marshall take the by-pass and turn right onto the N.C. 213 ramp to Mars Hill and then turn left. Near mile sixty-three, the route enters the city limits of Mars Hill and passes the campus of Mars Hill College, a liberal arts college founded in 1856.

Past the college, the route intersects with U.S. 19/23. Cross over U.S. 19/23 and turn right onto the ramp for U.S. 19/23 north toward Johnson City, Tennessee. Stay on U.S. 23 (in the future this route will be called I-26).

From this point to the Tennessee state line (a distance of about twelve miles), the route is under reconstruction by the North Carolina Department of Transportation. This section of new road (estimated to be completed in 2002) will connect with the new four-lane highway at the Tennessee state line, and when completed will be part of Interstate 26.

The North Carolina section of new construction will excavate approximately 30 million cubic yards of soil and rock to build the new interstate. The Tennessee section of mountain road (15 miles in length) required the excavation of 26 million cubic yards of soil and rock (Moore, 1997).

Stay on this old section of U.S. 23 until it crosses the top of the mountain, where it merges onto the new highway at the Tennessee–North Carolina border. This fifteen-mile, four-lane section of road cost Tennessee $170 million to build. In fact, enough dirt and rock were removed during the construction to fill the University of Tennessee's Neyland Stadium to the top twenty-one times.

The engineering geology factors that had to be considered in creating this stretch of road were monumental, including rockslope stability, rockfall, embankment stability, groundwater drainage, acid drainage, and foundation conditions for bridges and retaining walls. The construction of U.S. 23 (future I-26) through the mountainous area of Unicoi County, Tennessee, required the use of a special design for roadway embankments (commonly referred to as fills). The design concept of underbenching, in which small benches (ten to fifteen feet in width and height) are cut into the natural surface and repeated along the slope and then covered with roadway fill, was used on all the highway embankments on this fifteen-mile stretch of road. This underbenching procedure makes the roadway fill stable and relatively immune to landslides. Of course, this engineering feat cannot be seen because it is now covered up by the highway.

The construction of this highway also demanded special designs to deal with unique environmental issues, which included acid-producing rock, serrated-cut slopes, erosion control measures (temporary seeding of slopes, rip-rap rock-lined ditches, siltation dams), bioengineering (planting of sprigs of fast-growing trees along relocated creek channels and along cut slopes), and habitat-crossing boxes to permit wildlife to cross beneath the road. In addition, concrete bridges, mechanically stabilized earth walls (a type of retaining wall), a tied-back anchored retaining wall, and numerous rock-fall catchment fences were constructed on this section of roadway.

This interesting stretch of road soon arrives at one of the most beautiful scenes in the Southeast. The state of Tennessee has built a spectacular scenic overlook about seven miles from the junction of the state lines where the Appalachian Trail crosses the highway and wiggles its way up the other side of the mountain.

This area boasts Unakite rock, named for the Unaka Mountains and defined as a granitic rock composed of pink feldspars, green epidote, black hornblende, and a clear to milky white quartz. This rock, easily polished and coveted by rock hounds, is quite visible along the roadcut sections.

The main overlook is so big, its shape resembles a helicopter landing pad. There are three more landings that spider off of a nature trail, also built by the state.

The overlook offers 360 degrees of staggering views of Flint Mountain, Snake Den Ridge, Higgins Ridge, Chigger Ridge, Longarm Ridge, Sampson Mountain, Big Pine Mountain, Bald Mountain, all rolling out 180 degrees behind the overlook. At this point, the visitor is at about three thousand feet up at the large overlook, made of mountain stonewalls.

The rocks supporting these mountains are composed of gneiss, schist, quartzite, slate, sandstone, and shale (fig. 20), some of which are over one billion years old. They are referred to by geologists as basement rocks. Other rocks, the Chilhowee Group quartzite and shales, are as young as five hundred million years old.

About one billion years ago, the North American continent was rifting apart (the separation of a landmass by splitting or "ripping" along a weak place in the earth's crust). The material that eroded off these two great landmasses was deposited into the sea that existed between them and became sediment. The process of erosion and sedimentation along the ancient landmass boundary continued for some 300 to 500 million years, and then about 500 million years ago, the two separated landmasses began moving toward each other. One landmass was the future country of Africa and the other was North America.

Fig. 20.
Gneiss is one of the geologic rock types found near Sams Gap along U.S. 23/I-26. Some of these rocks are believed to be over one billion years in age.

Sediments in these ancient seas have been lithified (that is, turned into rock), deformed, pushed upward, and then folded and faulted by the collision of the two vast landmasses. The grinding of the rock masses helped to create pressure and heat that also metamorphosed the rocks into the quartzite, slate, schist, and gneiss that we see today, elements that form the skeleton and foundation for these mountains (fig. 21).

If you take the nature trail at the overlook, with its convenient benches spaced about a hundred feet apart, it gains another three hundred feet or so in elevation. The uphill pedestrian walk is worth the view at the top, where the panoramic scene and the brisk morning air will leave any observer breathless. This new overlook offers views that rival anything that can be seen in the Great Smoky Mountains National Park, the Blue Ridge Parkway, or any other place in the high mountains of the Southeast.

The elevation is so high at this point that shadows from the low puffs of cumulus clouds chase along the ridgetops of mountains, many of which are made of the same rock that formed the Chilhowee Mountains in and around Knoxville. And many of the mountains in this panoramic view are formed from rock more than one billion years old.

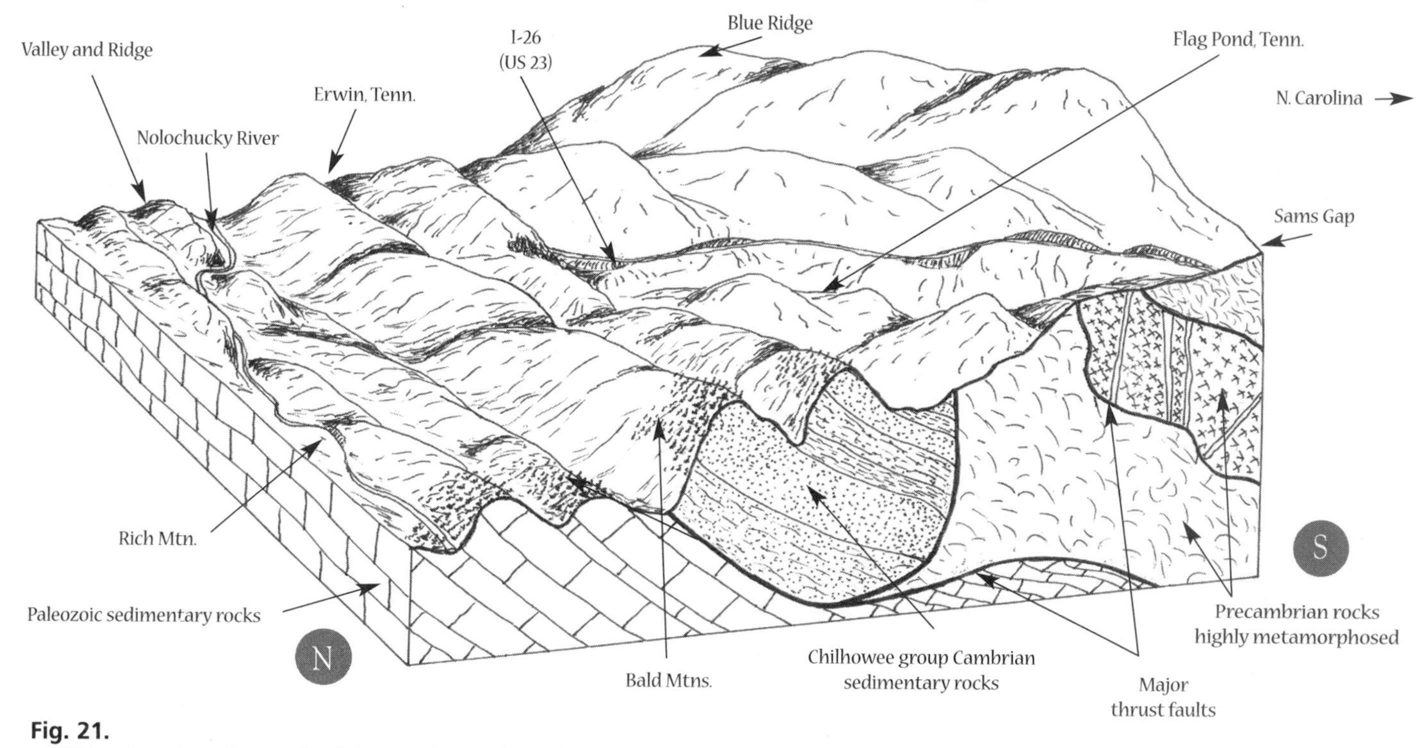

Fig. 21.
This schematic and generalized drawing depicts the geologic structure in the Flag Pond and Sams Gap section of Unicoi County, Tennessee, along U.S. 23/I-26.

After the overlook, stay on the four-lane until it reaches highway 81 and 107. Take Exit 18 at Erwin, Tennessee. Here, turn left onto State Route 81, which leads through a gorge of the Nolichucky River, where a sheer four-hundred-foot-high bluff of Chilhowee Group quartzite is exposed along the river. There are shoals or shallow water rapids in this gorge where bedrock is exposed in the riverbottom. The highway crosses the Nolichucky River, a very scenic area, shortly afterwards.

Erwin is in Unicoi County, one of the most interesting sections of upper East Tennessee. The county is shaped like a large bone, big on both ends and slender in the middle. It is about twenty-eight miles long and five miles wide, and the famed Nolichucky River runs in a northerly direction through the middle of the county.

Erwin is nestled in the valley between the Unaka Mountains and the Buffalo and Rich Mountain Ridges. Before white Europeans arrived, this was a Woodland Indian village. Unicoi County was part of Washington and Carter Counties in the late eighteenth century, a place rich in minerals, timber, and rivers.

Erwin also holds the distinction of being the only city in the nation where a full-grown elephant was lynched in 1916. The elephant, named Mary, was thirty years old and weighed over five tons. The town had long awaited the arrival of the Sharps Circus, on its way from Kingsport, but just prior to departure for Erwin Mary had killed her trainer, Walter "Red" Eldridge (Brown, 1990).

Since Mary was also believed to have killed others, when the circus train arrived in Erwin her owners decided it was time to dispatch Mary with all haste. But how do you kill a five-ton elephant? There was not a gun large enough in Erwin to do the job, but at this juncture in its history, Erwin was a thriving railroad center. So it was decided to haul Mary to the railyard and use equipment from the Clinchfield Railroad to do away with her. A crane was sent near the old train depot, but when the noose was put around Mary's neck, the chain broke once. The second attempt, however, was successful, and both contrary Mary and the town of Erwin were forever locked into notoriety: the story of the town that hanged an elephant has become famous. Only about five hundred people witnessed the deed that day, but over the years more than five thousand people have come forward to say they were present.

In this area, the abundance of iron ore and manganese were used by the earliest settlers to make shovels, plows, axes, and cooking utensils. Those first settlers moved into a place that came to be called Greasy Cove and Limestone Cove, two of the most beautiful and fertile spots in the state.

It is not hard to understand why pioneers settled in Unicoi County's Greasy Cove. More than two hundred years ago, unlimited natural resources abounded in the area—it featured rich soil, virgin forests, and creeks and riverbanks studded with the natural filter of canebrakes.

The cove's geographical boundaries include many coves, hollows, creeks, and gorges along the foot of Unaka Mountain. The Nolichucky River and Buffalo Creek meander gracefully along the shadow of Buffalo Mountain.

The trails made by buffalo cuts through the forests were first used by pioneers called long hunters. These fearless hunters ventured far and wide, sometimes leaving their families for months at a time, to search for new riches in the Tennessee wilderness in the 1760s. They crossed the mountains into the Holston, Watauga, and Nolichucky Valleys. Among these explorers were Daniel Boone and Jacob Brown, who was the first known permanent white settler to set up a "pole cabin" in what was then called Nolichucky country.

Greasy Cove was important for other historical reasons. Here, in 1781, Gen. John Sevier, Tennessee's first governor, mustered an army for a campaign against the Middle Towns of the Cherokee Indians in North Carolina. It marked the beginning of the end for Native Americans in East Tennessee.

And it is well established that Andrew Jackson crossed into Tennessee for the very first time by way of Iron Mountain Gap, through Limestone Cove and Greasy Cove, on his way to Jonesborough. That town was the county seat of Washington County, of which Unicoi County was then a part.

By 1788, the cove was notorious for fast horses and gambling. That year, Andrew Jackson challenged Col. Robert Love to a horse race. Jackson, the fiery Scotch-Irish, was a poor loser. When he lost the race, he called the colonel and his friends a "band of land pirates." The colonel responded by calling the future seventh president of the United States a "long, gangling, sorrel-topped soapstick." The two men had to be led away from each other, as both struggled to get at their pistols.

Greasy Cove has yet another distinction. It was the site of the early kick-off of the War of the Roses campaign between brothers Robert Love Taylor and Alfred Taylor in the 1886 race for Tennessee governor (West, 1998). Both Taylor brothers served as governors of Tennessee, but the real fireworks did not explode until the 1886 governor's race, when they opposed each other. Alf, the Republican, lost to Bob, the Democrat. The campaign became known as the War of the Roses, since the brothers campaigned together and were rather gentle to each other. In fact, their mother had warned them that they had better not get nasty, or they would have to deal with her after the election. So the War of the Roses was a peaceful, humorous affair.

Unicoi County is settled almost completely between the links of the Unaka Mountain range, extending along the North Carolina border like an uneven chain. To settle this wild and wonderful country, it took men and women of steel and stern wills to cross Iron Mountain on the county's southeastern shoulder.

Unicoi County's human and natural history are almost unparalleled in this state, and perhaps in the nation. The county is bordered by some of the oldest-known mountains in the world, with hillsides matted in a staggering variety of vegetation. Millions of years ago the glacier period transported Canadian Red Spruce, Fraser Fir, Yellow Birch, Mountain Ash, Balsam, and other trees and shrubs to this region. Life-and-death struggles took place against this scenic background. Inhabitants had to survive from day to day. The dangers included death by scalping from Indians, who resented being displaced from their lands, by attacks from wild animals, or by starvation. But the rich natural resources made the area attractive enough for the first white people coming across the mountains to remain and establish their homes and lives there.

The Nolichucky River has a Cherokee name meaning "spruce tree place" or "mad rushing waters." It began as Nana-Tsugu or Nula-Tsu'gu-yi. The white settler had no notion of how to pronounce those words, so the name came out as Nolichucky. Before the advent of crosscut saws, this was a paradise where trees grew to astounding girth, some measuring nearly twenty feet in circumference and living to be more than 375 years old.

About eight miles outside of Erwin, turn left onto highway 107. Here the route leaves the Blue Ridge Province and enters the Valley and Ridge, where the ridges and valleys are much lower in height and size. The drive along S.R. 107 features many farms where produce, such as tomatoes, watermelon, cantaloupes, beans, and squash, are raised for market. These farms are developed on the floodplain of the Nolichucky River, where floodwaters have deposited rich sediments washed down from the higher slopes of the mountainous terrain.

Continue on highway 107 toward Tusculum and then to Greeneville, where U.S. 11-E connects with I-81, which then junctures with I-40.

You have been in fall's "faule," where clouds in autumn cracked.

Cherohala

Destination: *Cherohala Skyway, scenic road crossing mountainous border between Tellico Plains, Tennessee, and Robbinsville, North Carolina.*

Route: *I-75, S.R. 68, S.R. 165, N.C. 143, U.S. 129 (optional Foothills Parkway) (includes Monroe and Blount Counties in Tennessee and Graham County, North Carolina).*

Cities and Towns: *Sweetwater, Tellico Plains, Maryville, Tennessee; Robbinsville, North Carolina.*

Trip length: *Approximately 194 miles round trip from downtown Knoxville (taking optional Foothills Parkway, distance is approximately 196 miles).*

Nature of the roads: *Four-lane interstate highway to two-lane state highways; some sections of road are in mountainous terrain and have steep, curvy sections; Cherohala Skyway is subject to closure during winter months.*

Special features: *Very scenic Cherohala Skyway roadway; contains numerous scenic overlooks and pull-offs; restrooms on Tennessee side; historic Tellico Plains; Chilhowee Dam and lake; Santeetlah Lake; Foothills Parkway; and western border of Great Smoky Mountains National Park.*

Get ready for surprises on the Cherohala Tour. Within its boundaries, the landscape rises gently from a low of nine hundred feet in elevation to a dizzying high of five thousand feet. At one point, a continuous seven-mile-long segment climbs more than one mile high into the clouds. On this trip the route traverses portions of rolling hills and valleys in the Valley and Ridge sections and soars into the mountains of the Blue Ridge in others, all in one day. Rocks on this trip vary from Cambrian and Ordovician-age specimens (from 570 to 400 million years old) to the Precambrian of the Blue Ridge, some of which have been around for more than one billion years. Ageless sentinels of another era, they are the story keepers of our natural history.

The Cherohala tour passes through rolling farmland and high mountain country, including the Cherokee and Nantahala National Forests in Tennessee and North Carolina. This area also played an important role in shaping the human history of East Tennessee. The Woodland Indians were the first people to live in the area, and much later came the Cherokee. Then white Europeans arrived. They found this country wild, wonderful, and full of life. Fish choked the rivers and streams. Waterfowl flocks numbered in the thousands. Buffalo and deer herds roamed the broad plains. This was paradise. And they stayed.

Tracing the Cherohala—which derives its name from the combination of the Cherokee National Forest with the Nantahala National Forest—will consume an entire day, so there is no need to hurry.

These mountains were shoved up—not just pushed, but violently shoved up—during that cataclysmic time when the North American continent and the African continent's plates were colliding into each other. Over several hundred million years, while the immense blocks of stone were propelled up and over each other, some actually tipped completely upside down. The dramatic scenery should bring to mind the land's ancient past. At one time here, great animals roamed freely. Tigers with foot-long tusks hunted, mastodons trumpeted, and birds with wingspans the size of small airplanes flew overhead, screeching. Much later, after these animals had long disappeared, human beings entered the picture.

For early peoples, living in the valleys was preferable, because they offered level ground, fertile soil, and flowing water. But some sought the solitude of the high mountains, where caves and rock shelters were ready-made places of safety.

From Knoxville, take Interstate 75 South to Sweetwater (map 4). The beginning of the trip will provide a warmup for the fall brilliance that will be visible later. Even the interstate median strip will be on fire with color during most of this trip. At thirty-one miles from Knoxville, the route crosses the Tennessee River Bridge. Although there is nowhere to stop on the bridge, it is possible to look down a bit in both directions for a colorful reflection of fall's face. Even the water seems to be ablaze when the colors are at their peak. Look for the reds of sumac and maple and the yellows of the tulip poplar.

The Tennessee River flows from the left to right on I-75 South. The large curves or bends in the river are known as meanders, places where the stream has made a broad, sweeping turn in its course, or where hard bedrock forces the river to alter its direction. Meanders are characteristic of old-age streams, such as the Mississippi River. The Tennessee River in East Tennessee has

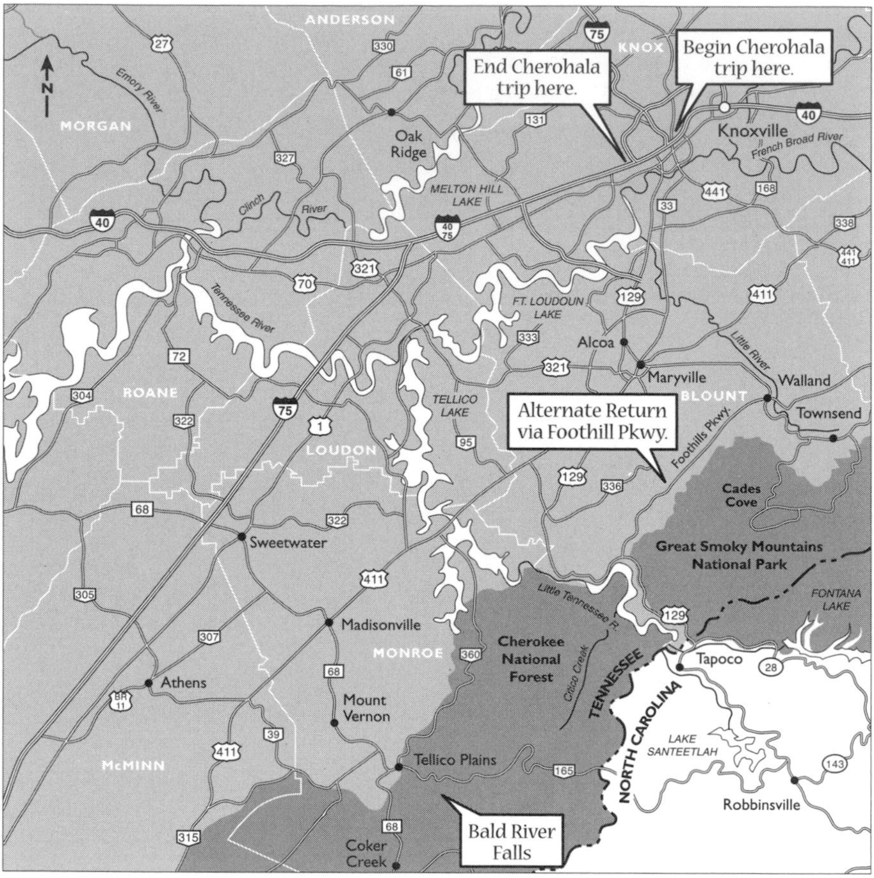

Map. 4.
The Cherohala trip.

some meanders in the river channel, but older streams consist of meander after meander, like a string of parentheses laid on their sides. In this sense, the Tennessee River is still young water, but its role in the history of upper East Tennessee is of great importance, for it served as the major highway for Native Americans, pathfinders, tradesmen, and pioneers. The Tennessee River system, the longest east of the Mississippi River, was extremely influential in the development of early settlement.

Once its lower waters were tamed, the Tennessee River became a river waterway, with steam-powered boats. But in the early days, canoes, flatboats, and rafts were the usual vessels of choice. These vessels were slow and ponderous, however, providing easy targets for the Cherokee, Chickasaw, Creek, and Chickamaugan Indians trying to hold off the tidal wave of white

Backbone Rock in Johnson County provides a unique opportunity to see very hard quartzite of the Chilhowee Group rock strata and the fall season makes it that much more enjoyable.

The fall colors brighten up the slopes and foothills of the Unicoi Mountains along U.S. 23/ I-26 in Unicoi County, Tennessee.

Maple trees provide some of the more brilliant colors of fall as these maple leaves show.

Hiking trails all over the East Tennessee mountains allow up-close views of the fall colors. This is the trail to Albright Cove near Cosby, Tennessee, in the Great Smoky Mountains National Park.

State Secondary Highway 133 in Johnson County winds through a stunning fall color display.

Rich Mountain Road winds across Cades Cove Mountain and Rich Mountain to connect Cades Cove and Townsend in the Great Smoky Mountains National Park.

Tulip poplars line an old roadbed in the Cosby section of the Smoky Mountains. Inset: The leaf of a beech tree floats by on an "Indian Summer" October day.

Tulip poplars often tall and straight, offer canopies of yellow during the fall color season.

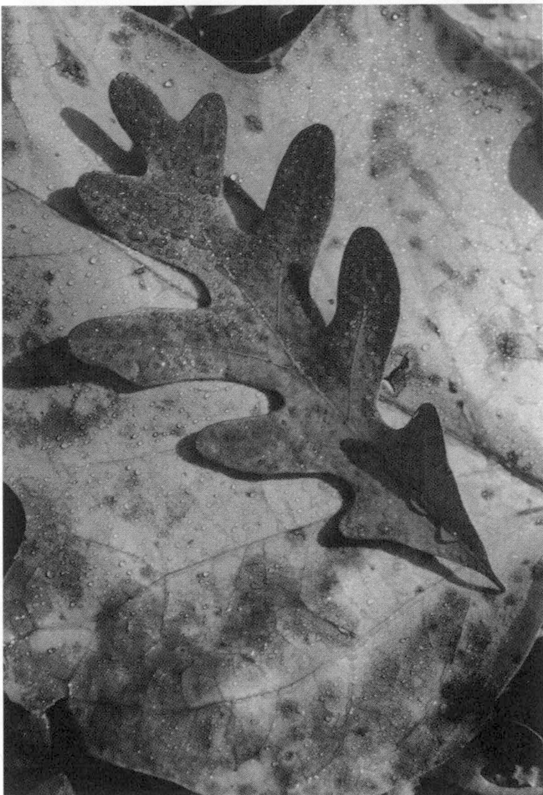

Above: State Route 66 winds across Clinch Mountain in Hawkins County presenting roadside views of the fall colors.

Left: Brownish colors of this oak tree leaf contrasts with the yellow of the tulip poplar.

The flowering dogwood leaf has its own shade of red that is accentuated by its very red berries.

The red maple displays very red and characteristic colors.

humanity on its way to take over their hunting grounds. The Native Americans were facing a formidable foe, one who could adapt and mimic, one who could sustain himself just as the Indian could, and one who was more determined to dig in and stay than the Indians were to drive them out. The white Europeans were dogged and determined, wily and skillful, also quite ruthless, and eventually far outnumbered the Indians.

After crossing the Tennessee River, I-75 follows a broad valley, which is part of the Valley and Ridge landscape. Be prepared for the spectacular panorama that comes into view while entering Monroe County.

From I-75, take the Sweetwater Exit onto highway 68. After turning left, heading toward Sweetwater, the route cuts across several ridges and valleys that indicate a typical geologic trend of the Valley and Ridge Province that makes up most of East Tennessee.

Evidence of the presence of the Cherokee Indian lives on in some of the old and some now-lost names of the towns of Monroe County: Chota, Citico, Toqua, and Tallequah. Lush mountains like Unicoi and Cataska carry the stamp of the Indian as well. Unaccustomed to Native American languages, pioneers added their own pronunciations to the language of the Indian. Soitee Woitee became Sweetwater. Tallequah became Tellico. Unaka became Unicoi (*Goodspeed's History of Tennessee*, 1887).

The Little Tennessee River and the incredibly beautiful Tellico River fall out of the mountains like streams of crystalline jewels. This was the domain of famous Cherokee chieftains like Hanging Maw, Water Hunter, Middlestriker, and Black Fox. The great Spanish explorer Hernando de Soto was awed by the land here when he ventured into it in the 1540s to hunt for gold.

This tour passes through Sweetwater, a town the sawmill built. And after the route crosses U.S. Highway 11, it will pass the Lost Sea, which is formed in rocks that were once part of a Bryozoan reef system. Bryozoa, which means "moss animal," are found in limestone and shale rocks Ordovician in age (roughly 440–500 million years old).

The limestone emanated from a sea-level ocean reef that extended tens of feet below the water, where the reef animal called Bryozoa built up on each successive layer, forming the actual structure of the reef. The sea was hundreds and hundreds of miles across. The pinkish gray limestone in evidence here is crystalline rock that has undergone solution activity by groundwater along numerous fractures in the strata. This means the rock has been dissolved by the groundwater, which is actually a mild form of carbonic acid, to create solution cavities in the rock that we call caves—a process geologists know as solution activity. Simply put, the slightly acidic groundwater infiltrates cracks in the folded and upturned limestone and then gradually dissolves the limestone material exposed along the crack surfaces.

In time, these cracks begin to gain in dimension and depth because of the increased amounts of groundwater, which then form numerous interconnected cavities. As these cavities open even wider, they act as a giant funnel, carrying the groundwater in ever-increasing amounts, which in turn enlarges the cracks even more. That is the precise formula for the creation of a cave.

The Lost Sea is just such a limestone cavern that has developed many, many numerous, massive cavities, some the size of several houses sitting side by side. Once known as Craighead Caverns (thought to be named after an American Indian), the Lost Sea has been a commercial cave since the early 1970s, but it is so old that the remains of a Pleistocene-aged jaguar, with a footprint six inches in diameter, have been found in the cave. The Pleistocene ranges from about 10,000 to 12,000 years ago to nearly 1,800,000 years ago.

The high ridges on this portion of the trip slant from northeast to southwest, creating hilly terrain. This rolling topography, caused by the underlying rock strata, continues to the town of Tellico Plains, where the region changes abruptly and becomes very mountainous.

Note the mammoth hills rising in front on the outskirts of Madisonville as the route heads toward Tellico Plains. Just as the road rounds the curve about sixty-nine miles from Knoxville, a breathtaking scene emerges—the Blue Ridge Mountains rising majestically in muted blues and greens. But in the fall, the mountains change their quiet hues for riotous autumn colors. Rocks in this section of the route are mainly Cambrian and Ordovician-age limestone, sandstone, and shale.

About seventy miles from Knoxville, turn left onto highway 165. This route passes through downtown Tellico Plains, a town created by the sawmill. This was, and still somewhat resembles, a frontier town with streets of mud and wooden sidewalks. In those days, timber floated like toothpicks down the Tellico River to the waiting sawmills in Tellico Plains. Entire mountains of trees disappeared, food for the young and hungry industry. And when it was over, the town had become a plains and the mountains were stripped bare. Today there are second-, third-, and fourth-generation trees, making a comeback in the forests, which are still managed for timber and the saw.

At seventy-one miles, continue on highway 165 as it veers into Cherohala Skyway, where something unusual takes place. The trip crosses a major geologic and physiographic boundary (fig. 22). In geologic terms, a fault surfaces from the deep like a great whale coming up for air. The geologically extensive Great Smoky Fault breaches here, bringing extremely old Precambrian-age rocks into contact with younger Cambrian- and Ordovician-age rocks. At first, very crumpled grayish black, shaley-looking strata are visible. These rocks are actually phyllite and slate, a thin rock that has been heated and subjected to tremendous pressure and metamorphosed. These rocks will also appear rusty looking, the result of the oxidation of the mineral pyrite, an iron disulfide mineral commonly referred to as "fool's gold."

Where there is fool's gold, there is usually real gold. In the upper sections of Monroe County, gold embedded in quartz was discovered in Coker Creek. The presence of quartz in Coker Creek can be explained by two cataclysmic events that took place, millions of years apart, beginning some 300 to 600 million years ago. The great geologic drama was staged on a gigantic sea and a single, vast mother continent from which all future continents were spawned. Like children, parts of the continent drifted away, creating oceans and seas, plains and mountains, basins and ranges, and mountain chains.

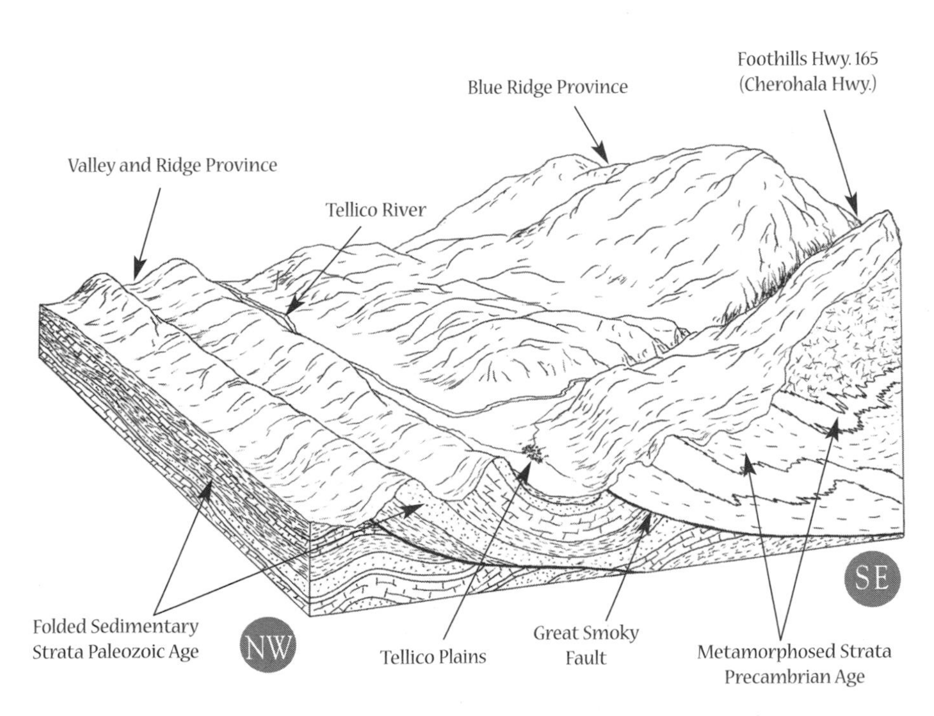

Fig. 22.
The Cherohala Parkway traverses the Blue Ridge Province as State Highway 165 leaves the Valley and Ridge as shown in this very generalized drawing.

The event occurred when the tectonic plates of what is now Africa collided with what is now North America, pushing up the mountains we now know as the Blue Ridge and the Appalachian Mountains and forming a gigantic landmass called Pangaea (Tarbuck and Lutgens, 1996). The mountains were folded, buckled, wrapped, and warped into existence. Once the mountain building began, over millions of years, the Unaka Mountains formed as the deep ocean sediments were thrust westward, and as the super continent broke apart, the Atlantic Ocean formed. During these epochal events, belts of rocks and minerals were scattered in huge, continental layers across wide regions of the world, and gold was literally formed in quartz veins inside the mountains throughout the eastern United States from northern Georgia to Tennessee and North Carolina.

Gold was first discovered in the late 1790s in North Carolina and in the mid-1820s in Coker Creek, long before anyone had heard of Sutter's Mill in California, where the gold rush occurred in 1848. The discovery of gold in Coker Creek changed that small mountain community and region forever, and eventually, because of the influx of prospecting settlers, played a significant role in the forced removal of the Cherokee Indians from their native lands to Oklahoma in the West.

The gold deposit in Coker Creek is known as a placer deposit. The erosion of the "mother vein," or original ore body, and the deposition of those gold-containing sediments in the floodplain deposits of a stream form placer deposits. The stream in this case is Coker Creek, and the original location of the gold deposit is unknown, but is believed to have been high up in the Blue Ridge Mountains, in a place possibly eroded and nonexistent today (Brown, 1991).

Coinciding with the location of gold placer deposits in the mountains, the journey leaves the Valley and Ridge Province and enters the Blue Ridge Province, which is marked by an abrupt physical change from low valleys and ridges to high mountain country.

A rainbow of color envelopes the route as it snakes along the Tellico River. Then, at seventy-three miles from Knoxville, it enters the Cherokee National Forest. A few miles inside the National Forest there is a turnoff to Bald River Falls (fig. 23). Take it! This side trip is a beautiful seven-mile drive. Numerous red maples, hickories, dogwood, ironwood, beech, redgum and black cherry paint this stretch of road for the visitor. Retrace the route, back to State Road 165, where the tour turns right onto State Road 165 and then to the North Carolina state line.

Stay on highway 165. The route will run past metamorphic rock that was deposited in the bottom of a deep ocean basin. The rock is grayish in color

Fig. 23.
Along the Cherohala Parkway near Tellico Plains there is a short seven-mile side trip to Bald River Falls, which is especially beautiful during the fall season.

with the rusty stain of pyrite. When these rocks were still soft sediments, they lay deep in the sea, possibly several thousand feet below the surface, in a submarine basin. Oxygen was almost nonexistent at this depth, and as a result sulfide minerals, such as pyrite and chalcopyrite, were created in the ooze of the sediment. When exposed to the oxygen and water of present-day earth, the minerals oxidized, giving the rock surface its rusty color. This trip gives you the full flavor of fall in East Tennessee and its varied geology—from old age to rust, from sea bottom to faulted rocks rising into the air.

Watch closely on the next leg, because you must make certain *not* to turn off at the intersection of Indian Boundary. Continue on Tennessee Highway 165, which bears to the right. After passing the Indian Boundary camping area, the route climbs, heading toward the skyline. Prepare for some of the most spectacular scenery to be found in this region. The Skyway is aptly named. This beautiful road is not without its problems, and the underlying rock is the culprit. It contains pyrite, a disulfide mineral that causes rainwater to turn into sulfuric acid before running down the mountain. This fool's gold deposit can produce quite a mess in such a vegetative landscape. When the road was built in the early 1970s, excavation of the grayish black slate rock exposed strata that contained the brass-colored mineral. These excavated rocks were placed in the roadway as fill to form the road. Even before the road was completed, acid drainage was already affecting local streams.

Dr. Don Byerly, a geologist from the University of Tennessee, pioneered a method to repair the acid-drainage problem along roadways (Byerly, 1981). He assessed the acid rock condition and began studies to correct the damage already done and to evaluate the location of the new road ahead into North Carolina (the road you will travel on this trip). The correction involved methods of adding lime and limestone rock to the effected pyritic rock in the roadbed.

The views off in the distance to the north and east are unbelievably beautiful. Pull off (to the left) at eighty-nine miles from Knoxville for a panoramic view of the foothills in the Blue Ridge. The view is across the Citico Creek Wilderness. There is the faint shadow of Cowcamp Ridge, and, a way off in the blur, Salt Spring Mountain. These are mountains in interconnected rows, where curve after curve brings a new scene and a new ruffling of uplands, like waves in the sky. The scenery brings to mind the ring of the axe and the scrape of the adz that were used to develop this area. Listen to the wind, and the faintest hint of fiddle music may drift through.

At ninety-three miles from Knoxville, the route will be up to 4,000 feet. Here beech and birch thrive and flaunt their gorgeous autumn gold. A mile farther on, the road crosses into North Carolina, about two hours from the start of the trip. Here Beech Gap rises 4,490 feet above sea level, and six miles later, the route climbs past Whigg Cove, where the elevation is at 5,000 feet. Be sure to catch the views from this vantage point. In this stretch of road, the route stays above 5,000 feet for about seven miles.

As the trip descends from the clouds, stop at Obadiah Gap pullout, which is about 108 miles from Knoxville. This is one of the best views on the entire trip. The waltz of the colorful trees in the wind provides a stunning sight. It seems as if there is a wind spirit at work here. Then just peer out into the hills and these unconquerable mountains. This is a view across the Santeetlah Wildlife Management Area, a scene whose beauty will last in memory throughout the long winter months.

In this area, the landscape reflects the erosion-resistant rocks of metamorphic origin and the Precambrian age, rocks that are mostly ocean sediments that have been metamorphosed as a result of the plate tectonics that brought the two continents of Africa and North America crashing together. As you move farther to the southeast, the degree of change in the rock (metamorphism) becomes greater. The degree of metamorphism can usually be traced to the type of minerals found in the rock. For example, the lower degrees of metamorphism contain the mineral chlorite (a greenish member of the mica group). The next zone is characterized by the presence of Biotite mica (the black variety of mica). As heat and pressure increased over time, the formation of garnets took place, marking the next higher degree of metamorphism. Rocks in this area sometimes contain small garnets—darkish red, pinhead-sized minerals in the rock—although most of them are not usually considered to be of gem quality.

Right now, the route is on highway 143. At 112 miles, it leaves Nantahala National Forest. The road becomes a repetition of curves once it returns to Tennessee. At 119 miles, turn left onto highway 143 East. This will lead to U.S. Highway 129. Turn left and head back toward Maryville, Tennessee.

As the route crosses back into Tennessee and travels along Chilhowee Lake, pay attention to the rock-cut exposures on the right. Some of these cuts reveal extremely tightly folded strata, called "kink" folds or chevron folds because of their tight repetitive "zigzag" structure. This folding is the result of the immense pressure these sediments and strata were placed under many millions of years ago.

Pass Chilhowee Dam, and at the end of Chilhowee Mountain the route crosses over the Great Smoky Fault, which demarcates the boundary between the Blue Ridge and the Valley and Ridge provinces. From this point on to Maryville and back to Knoxville, the trip will be in the Valley and Ridge. Here the rocks are younger (from the Paleozoic age) and are sedimentary in origin.

An alternate route back to Maryville and then on to Knoxville is to take the Foothills Parkway from U.S. 129 over to Walland. This dramatic roadway glides along the crest of Chilhowee Mountain, offering spectacular views of the Great Smoky Mountains on one side of the mountain and vistas of the Valley and Ridge on the other side.

A must stop is the observation tower at Look Rock, located along the crest of the mountain about seven miles from U.S. 129. Fantastic photographs can be taken from here.

At Walland, exit the Foothills Parkway and turn left onto U.S. 321 and head toward Maryville and then back to Knoxville on U.S. Highway 129.

From the time this tour leaves highway 68, it is in Nantahala National Forest, and the speed limit is forty-five miles per hour. You will want to take your time anyway, regardless of the speed limit: the sensational vistas are lit up so brilliantly at this time of year that each deserves to be savored.

Rich Mountain

Destination: *Rich Mountain Road (one-way gravel road between Cades Cove and Townsend).*

Route: *Begins in Maryville at intersection of U.S. 321 and S.R. 73; S.R. 73 to Great Smoky Mountain National Park Townsend entrance; Park roads–Laurel Creek Road, Loop Road, Rich Mountain Road (entire trip in Blount County).*

Cities and towns: *Maryville, Walland, Townsend.*

Trip length: *Approximately sixty miles roundtrip.*

Nature of the roads: *Two-lane paved state highways, two-lane paved park roads, one way one-lane paved and gravel park roads; one-lane Rich Mountain Road is gravel and one way—it is also closed daily at sunset and generally closed during the winter months; park speed limits are much lower than state highways.*

Special features: *Historic Maryville and Maryville College, Townsend, Great Smoky Mountain National Park (including Cades Cove), very scenic Rich Mountain Road.*

On the Rich Mountain trip, the geology and landforms will convey a true sense of what it was like to be a mountain person, living according to the varying rhythms of mountain life. The tour evokes the voices of old mountain folk, the ringing of the mountain fiddle, the whispering whisk of the mountain whippoorwill flitting through the trees, or the haunting, far-off sound of an axe chopping wood. On this trip, the region's history is revealed, a time when isolated families lived and worked in an agrarian society, completely self-sustaining. The mountains leading into Cades Cove, which is featured near the end of this excursion, were once home to folks who raised crops in a seasonal cadence and lived in harmony with neighbors. They cut identifying notches in the ears of their cattle and let them graze in open ranges. An animal's notched ear represented a family's special mark, and if a cow strayed, a neighbor knew by the notch where the animal belonged. Sometimes in the summer, farmers moved their cattle to the high mountain balds, where wild grasses grow, and brought them down the mountain again in the fall. It was a time when people took their promises as their bond, and land and life were sacred.

The geographic features of the roads of this trip include limestone coves, Valley and Ridge and Blue Ridge topography, "geologic windows," and mountain slopes. The varied rock exposures include deformed shale, quartzite, limestone, phyllite, and conglomerate. The trip begins in the rolling Valley and Ridge Province and enters the mountainous Blue Ridge Province just south of Maryville, Tennessee. Those blue ridges ripple like waves in the sky, riding out from Georgia, across Tennessee, North Carolina, Virginia, and on into Maine.

Before taking this trip, gaining a geologic perspective of the Great Smoky Mountains may be helpful. Find a spot in the Great Smoky Mountains National Park where Mount LeConte can be clearly seen. A good place to do this is near "Fighting Creek Gap" along Little River Road, just up from the Sugarlands Visitor Center. Find an overlook near the gap and look back down the valley toward the Visitor Center. Mount LeConte will be to the right, high up on the skyline. The ridges hanging above make up a thin, distant smudge from the top of Mount LeConte that is a trail in time. The ridgeline fades northerly in a long decline through a blue haze, like a pencil mark being erased, then skips across the sawtooth ridges of Webb Mountain, Shields Mountain, Bear Wallow Mountain, and on into the beyond.

Like an ancient traveler resting now, the sloping remnant of ridgeline remains in beautiful isolation, defining our history in a language of stones from the dawn of time. That line hovering in the sky is a visible reminder of an ancestral floor that supported this region two to twenty million yeas ago. Now we live below the ancient floor, riding an erosional elevator to the bottom. We exist on a floor formed on rock that moved more than two hundred miles from the southeast to the northeast from what is present-day Columbia, South Carolina, and Charlotte, North Carolina, both of which were under an ancient ocean. When the continents collided approximately 230 million years ago, so much energy was created that rocks that had been on the bottom of an ocean floor were pushed up and over younger rock and then moved northwest. Then, over millions more years, the continents began pulling apart, producing our present-day mountains. Colossal upheavals of crashing continents over hundreds of millions of years pushed up mountains, valleys, and ridges from the bottom of that ocean. We began

below sea level, and now Clingman's Dome is 6,643 feet above its origins, riding a slow-moving thrust fault to the top and, over several hundred million years, punching up through the landscape. East Tennessee was part of the Alleghanian Orogeny, or mountain building, in the heavy geologic action that occurred over millions and millions of years.

Cades Cove is part of the floor that continues its slow descent through time. One day, millions of years from now, Cades Cove and Townsend will merge to form a single larger cove as erosion proceeds to wear away Cades Cove Mountain and Rich Mountain. To learn about the past, geologists look at the present world to see how events unfolded millions of years ago. It's an idea first developed in the late 1700s by James Hutton, a Scottish physician, farmer, and the father of modern geology. This basic geologic principle is referred to as uniformitarianism (Tarbuck and Lutgens, 1996). By observing these "geologic windows"—seeing the past and future by analyzing present geologic conditions—it's possible to understand a great deal about the earth around us. On this trip there are outstanding examples of such geologic windows, just one of the unique features found on this fall trip, in Townsend and Cades Cove.

The Rich Mountain trip begins in Maryville, Tennessee (map 5), just past Alcoa, home to three large plants belonging to the Aluminum Company of America. Bauxite, the ore of aluminum, is shipped by rail to this area, where it is smelted and refined into aluminum. Alcoa chose to build its plant in this area because of the abundant source of cheap hydroelectric power. A number of the dams located on the Little Tennessee River were originally built by the Alcoa Company.

Travel the brief distance from Alcoa to Maryville at the juncture of U.S. 321 and State Highway 73. At the beginning of the trip, the route will pass the campus of Maryville College, one of the oldest educational institutions in the region and state. Founded in 1819 by Scotch-Irish Presbyterian minister Isaac Anderson, the college is run by the Presbyterian Church. Just beyond the college on the right is the historic Thompson-Brown House (c. 1800), one of the oldest two-story log structures left standing in East Tennessee (Blount County Historic Trust, 1991). The building is of particular interest to archaeologists because it is made of hewn pine logs chinked with plaster in a mixture of animal hair. Continue south on highway 73 and U.S. 321.

Just outside Maryville, the Great Smoky Mountains begin to emerge on the horizon, beginning with Three Sisters Mountains on Chilhowee Mountain. At 1.6 miles these formations cut the horizon in three perfectly shaped peaks. Formed by the erosion of steeply dipping rock strata of the Chilhowee

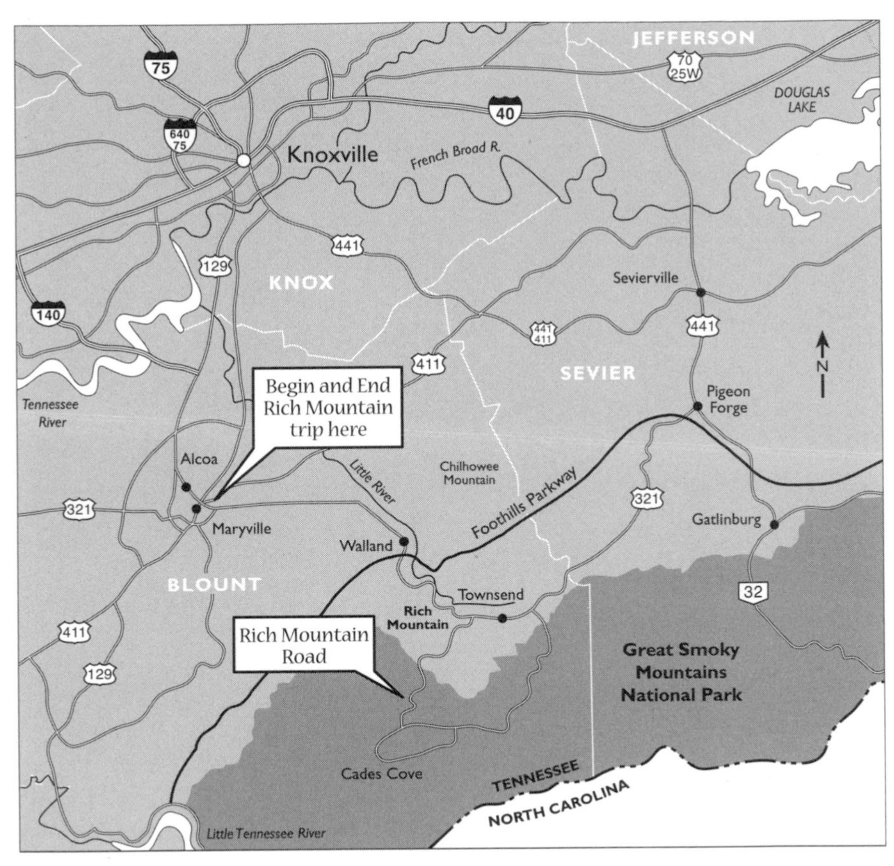

Map 5.
The Rich Mountain trip.

Group, the mountains are composed of shales and quartzite and mark the beginning of the Blue Ridge Mountains.

The route continues across tilted strata of Cambrian- and Ordovician-age limestone and shale. Most of the bedrock is covered with reddish brown clay soil, the result of many thousands of years of weathering and erosion. At log mile 5.8 is a dramatic view of Chilhowee Mountain, which marks the boundary between the Valley and Ridge and the Blue Ridge. The closer the route gets to the boundary zone, the more intensely folded and fractured the rocks become. Along the road, the brownish gray exposures of shale and the gray, rounded limestone masses protruding in the grassy pastures become more deformed on the approach to Chilhowee Mountain.

About 8.5 miles from the beginning of the trip is highly deformed and crumpled shale strata along the right side of the road as the route enters the gap in Chilhowee Mountain. This is Ordovician-age shale that has been folded into very tight bands by the movement of older rocks in the Great Smoky Fault, a fault plane that exists not just beneath Chilhowee Mountain, but beneath the Great Smoky Mountain National Park as well.

In this area there are farmers on tractors in the fields, gathering hay for winter. Although the family farm has long since passed into extinction in the region, some families continue to live on the land in this area, keeping cattle and hogs. Once, farming was a way of life for the people here, but now only an occasional farmer on a tractor can be spotted, working his sundown job.

Just past these crumpled beds of shale are upturned beds of Chilhowee Group sandstone. These once flat-lying sedimentary layers were bent upward by the compression of the colliding continents, exposing the Great Smoky Fault. Above this spot (near log mile 8.9) is a checkerboard-patterned, concrete retaining wall called a tied-back wall, made of a series of concrete panels anchored to the bedrock that exists beneath the mountain. This type of wall was constructed to keep from excavating the side of the mountain where unstable deposits exist.

The movement of the rock strata by thrust faulting is the result of compression forces produced by the gradual collision of the North American continent with the African continent as long as 230 to 250 million years ago. This convergence, which produced the Appalachian mountain-building episode, is known to geologists as the Alleghanian Orogeny. The Foothills Parkway is on the right at 9.7 miles. The road cut just past the parkway interchange exposes medium-gray beds of dolostone that date from the Cambrian-age Shady Formation, and a mile or so past the Parkway there is Precambrian-age rock. With road construction under way, these roads are changing, and soon some of these roadside hills will be a thing of the past. The rocks are phylite and slate, and in the autumn the hillsides seem to be aflame with deep reds and bright yellows. Sourwood, which is almost synonymous with the romance of the mountains, and blackgum brightly color the hills.

For the next several miles, exposures of Precambrian-age strata exhibit the extreme folding that resulted from the collision of the continents. Near mile 12.9 there are very tightly compacted rocks that illustrate chevron folding, a tight zigzag pattern resembling the stripes of an army officer's insignia. These folds can be seen at the cut slope alongside side the roadway, but be careful, because there are no shoulders along the road and traffic can be fast and heavy.

At thirteen-plus miles, the route begins to enter the cove country, which geologists call a "geologic window." The landscape opens up, broadens into a form of a cup, and spreads out, as if reclining for a long nap. The first of the coves is Tuckaleechee Cove, which contains Dry Valley, a valley that got its name because all the surface water that falls on the land in this valley flows into subsurface caves, resulting in an absence of surface creeks and streams. The Indians indicated the absence of water by saying that it never rained here.

Tuckaleechee Cove (Townsend) is an excellent place to view the Great Smoky thrust fault, where erosion has cut down through the older overlying Precambrian rocks and exposed the more soluble limestone and shale of the younger Paleozoic age. Tuckaleechee Cove is a geologic window where the floor of the cove is younger rock, surrounded by mountainsides of overlying older rock. In places like this, the observer looks through the "window" of older rock to see the younger rock.

Because of the soluble nature of limestone bedrock, there are many caves in Tuckaleechee Cove. The commercialized Tuckaleechee Caverns are the most widely known of the caves in the Townsend area. Over twenty caves have been found and explored in the cove.

On highway 321, you are now only a short distance from the National Park. Color surrounds the road as it approach the little village of Townsend, which began in 1898 as a logging camp. Dr. W. B. Townsend of Clearfield, Pennsylvania, bought one hundred thousand acres of this ancient land in Tuckaleechee Cove and began cutting the old-growth trees, creating the first of many logging and timbering operations that exploited the virgin forests of this region. Later came the destructive tanneries that used tannin from the trees to produce leather. When the trees were gone, so too was Townsend, leaving behind a ravaged countryside that has finally returned to second- and third-generation trees. And now the little village, once the hunting grounds of the Cherokees and other Native American tribes in this region, is being opened up to a five-lane highway system by the state of Tennessee. This will undoubtedly change Townsend once again, as tourists find that the "Quiet Side of the Smokies" has now become just like the other side of the Smokies, with its garish billboards and dozens of retail outlets that sprawl in uncontrolled growth across the face of Gatlinburg and Pigeon Forge. Townsend is becoming so commercialized that entrepreneurs are selling bonsai trees out of tents, and tourists can purchase a cell phone and a pager at the same place they pick up bags of boiled peanuts.

At 16.9 miles, highway 321 cuts left toward Wears Valley and Pigeon Forge. Stay on S.R. 73 into the Great Smoky Mountains National Park. About a half mile past the U.S. 321 turnoff, the road passes Cedar Bluff, a

Fig. 24.
In this view looking north from S.R. 73, Cedar Bluff in Townsend displays folded and faulted Paleozoic rocks along the Little River and State Highway 73.

huge gray limestone bluff easily visible on the left side of the highway, across Little River (fig. 24). The bluff is medium- to thick-bedded limestone and dolostone strata of the Paleozoic-age Knox Group, with numerous faults in the bluff face formed some 230 million years ago. The dipping layers of rock that abruptly change their orientation, creating a linear surface that separates the rock layers, constitute a fault exposure. The Cedar Bluff exposure was fundamental in helping geologists unravel the history of the Great Smoky Fault. The fault is best viewed in winter when vegetation has died. In the fall, the bluff is festooned with fall colors.

Just down from the boiled peanuts and cell-phone outlets, this tour once again crosses over the Smoky Mountain thrust sheet, passing through a geologic wrinkle. The route goes by an exposure of rock along the right side of the road that shows the structure of the fault. Here strata of the Precambrian-age Metcalf Phyllite is thrust over the younger limestone and dolostone of the Paleozoic-age Knox Group. The limestone is gray and somewhat massively bedded. The phyllite is also gray and varies to brown, but is very thin bedded and has a shaley appearance.

By the time the route enters the Great Smoky Mountains National Park, much older rock predominates once again. At 19.3 miles, the road come to the Wye intersection, about 0.9 miles from the entrance of the park, where Little River meets Laurel Creek. Turn right onto Laurel Creek Road going toward Cades Cove. The sign points to Tremont, but pass the turnoff to Tremont at 19.7 miles and continue towards Cades Cove. Along this section of road, there are numerous exposures of Metcalf Phyllite. The thin, platy layering in the rock is not bedding, but foliation and cleavage resulting from the flexing and deformation induced by faulting and folding.

On the way to Cades Cove, take a look at Little River to the right. Millions of years ago, these ancestral waters actually began high above, and today the river continues to cut through metamorphic rock. Now the route is actually in a canyon that continues to widen as the river is diminished. Here the trees are reflected in the water, providing twice the thrill of all these autumn colors.

Approximately four miles from the Wye toward Cades Cove is the trailhead for a special hiking trail to Whiteoak Sinks (this is about 23.6 miles from the start of the trip). It is marked as the Schoolhouse Gap trail, which was once an old roadbed from the early settler days. Schoolhouse Gap, which is located at the park boundary, is flanked to the south by an enormous sinkhole area known as Whiteoak Sinks.

Whiteoak Sinks is an unusual place. It is an example of a small-scale geologic window. As mentioned before, areas where the younger Paleozoic limestone and shale have been exposed by weathering have formed mountain coves like Cades Cove and Tuckaleechee Cove. Erosion of the mountain ridges has connected Whiteoak Sinks with Tuckaleechee Cove. Topographically it is still a cove, but geologically it is a part of Tuckaleechee Cove. These coves are surrounded by older and topographically higher rock strata, which form "windows" down into the younger limestone strata of the cove. The deep, fertile soils produced by weathering limestone and the surrounding mountain rock have made the coves attractive places for humans to settle and farm. That is why Native Americans and the early pioneers hunted, settled, cleared, and farmed these areas.

The debris that has washed in from the surrounding slopes and covered the floor of Whiteoak Sinks consists of organic material and small broken pieces of bedrock, including silt and sand particles. This fertile soil mixture now supports a variety of beautiful wildflowers.

In these limestone coves, erosion of the bedrock has produced karst features such as sinkholes, caves, and subsurface drainage. Chemical weathering of the limestone bedrock, along with the groundwater percolating down through the rock, has enlarged fractures in the rock mass to produce a number of interconnecting solution channels that we know as caves. The cavernous bedrock is reflected at the surface as sinkholes and depressions. Continued erosion will enlarge the sinkholes so that they coalesce into very large areas of internal drainage, commonly referred to as karst valleys.

At least four such limestone caves are found in the park: Gregorys Cave (in Cades Cove), Bull Cave (along Rich Mountain Road), and Rainbow Cave and Blowing Cave (both in Whiteoak Sinks). Those who enjoy spelunking must obtain a permit from the superintendent's office of the Great Smoky Mountains National Park before entering any cave in the park.

Whiteoak Sinks is roughly twenty-five acres in size and is a drain for several hundred acres of mountain slopes. A number of families once lived in Whiteoak Sinks, which is also called Whiteoak Flats by some old-timers. Old stone fences, house foundations, pieces of pottery and glass, and even a gravesite can be found in the sink area. The families who lived there were forced to leave the area to make way for the formation of the national park some six decades ago, but they held on until the last minute, leaving food cooking on the stove and dishes in the sink.

The trail to Whiteoak Sinks begins at the dirt road at the parking area along Laurel Creek Road. This road will pass numerous trees in fall array as well as rhododendron and hemlock trees. Schoolhouse Gap is a two-mile uphill walk. The official trail ends at the park boundary.

Because the trail down into the sink area is not marked, it is difficult to find and follow. The trail to the sink follows a narrow footpath down into a rhododendron-covered ravine, crossing and recrossing the small branch several times (there are no foot bridges). About a third of a mile into this small ravine, a trail cuts right up over a saddle in the hill and back down into the sink proper. This was the original wagon road into this section of the sinks.

As the trail continues down into the sink, outcroppings of limestone strata indicate that you have crossed over the Great Smoky Fault and are in the geologic window. To the right and near the bottom of the sink is a very high limestone bluff. At the foot of this bluff is the entrance to Blowing Cave.

In the summer and fall of 1997, a gate was constructed at the entrance to Blowing Cave to protect unaware park visitors from falling into the cave and also to protect the bats that use the cave from interloping humans. During the summer months, a strong, continuous blast of cool cave air issues from the mouth of Blowing Cave and can offer a refreshing break from hot and humid weather. This is also a nice spot in the fall season. The golds and yellows of the tulip poplar and buckeye leaves yield a rich hue as light enters the sinkhole area.

Rainbow Cave is located in another section of Whiteoak Sinks, where a stream flows over a small bluff forming a waterfall into the mouth of a cave below. In the afternoon, the sun hits the waterfall and forms a small rainbow in the cave entrance. Behind the waterfall and along the rock bluff is the fault trace of the Great Smoky Fault, exposed as a dark, almost horizontal boundary line between the overlying older Precambrian rocks and the underlying Cambrian and Ordovician limestone that form the cave. This is a special place, because exposures of the Great Smoky Fault are rare, and most visitors never get to see one.

Geologists suspect that the water that flows into the subsurface drains through the cavern system and emerges at several springs down in the Dry Valley section of Tuckaleechee Cove (Townsend). The water falls a total of 550 feet from Whiteoak Sinks to the resurgence at Dunn Springs in Townsend, near Tuckaleechee Caverns.

Back on the road and heading toward Cades Cove, the elevation increases on the drive up Laurel Creek Road to Crib Gap. But get ready to drop down into Cades Cove, another geologic window like Tuckaleechee Cove in Townsend. At log mile 25.4, and about 6.7 miles from the Wye intersection, the road crosses Crib Gap, where it descends into Cades Cove (fig. 25).

As a result of the erosion, the valley is surrounded by ridges and mountains composed of older, faulted strata, while the valley floor is underlain by younger limestone and shale. During the last hundred thousand years or more, debris has been eroded and transported down the mountainsides by landslides and debris flows, water-saturated soil, and rock material, which becomes unstable and comes down the mountainside. This debris material now forms lobes of vegetated and elevated land that jut out onto the cove floor along the valley boundaries.

The ascent up the gap is marked by an enclosed yellow canopy that mixes with the deep evergreen shades of hemlock and white pine. Poplar, with its broad yellow leaf, and maples and hickories provide a colorful umbrella of fall color. Note also the Umbrella Magnolia and the beautiful Mountain Maple, which become translucent yellow when backlit by the sun.

At 26.7 miles is the Cades Cove campground on the left. The Loop Road, which begins at 27.0, is always crowded in the fall, so be prepared to go slowly. It's a one-way road and 11 miles around. During fall the entire 11 miles might be a traffic jam, but our route turns off to go across Cades Cove Mountain and Rich Mountain, leaving most of that traffic behind.

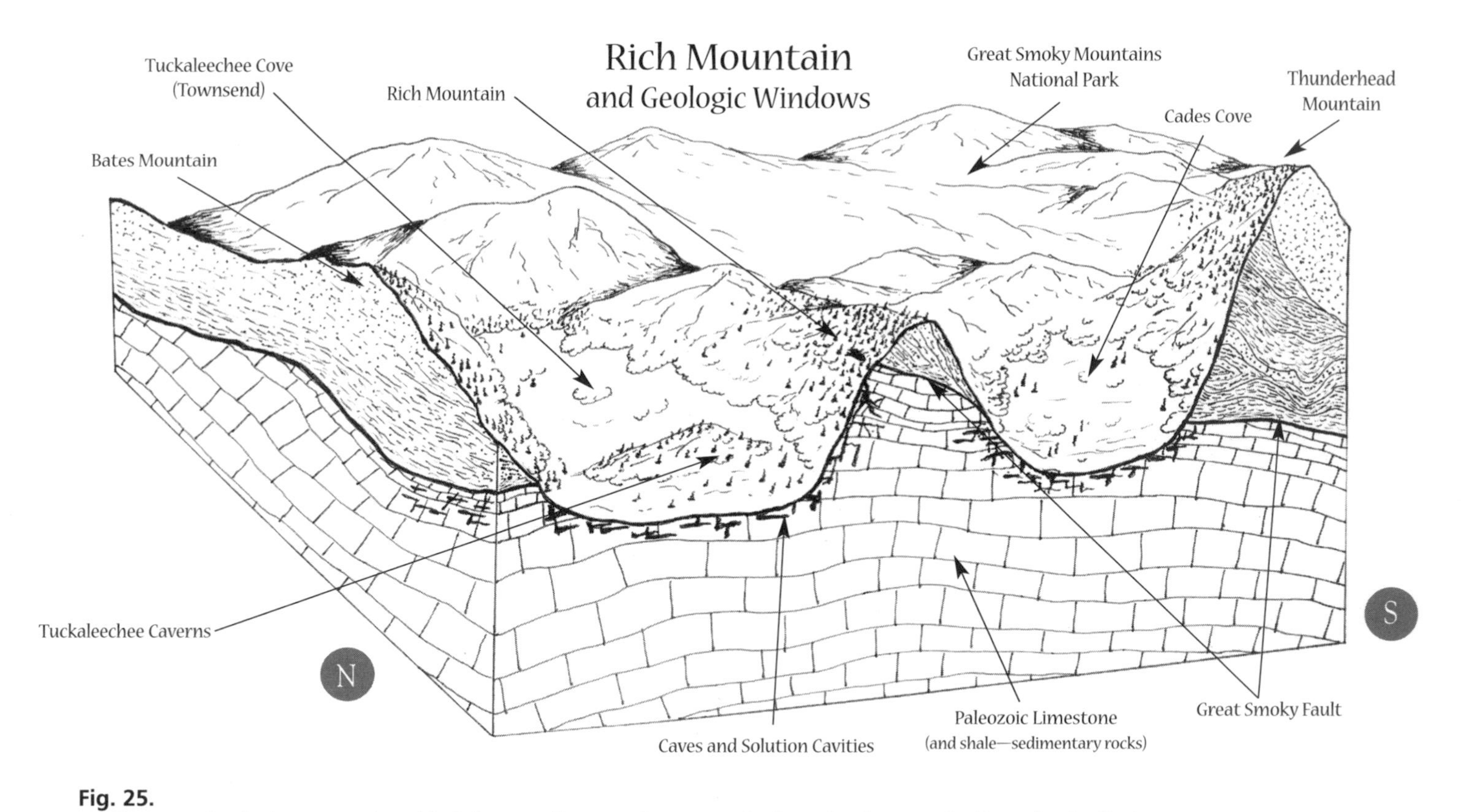

Fig. 25.
This generalized schematic geologic block diagram illustrates the structural relationship between the Great Smoky Fault and the formation of geologic windows such as Cades Cove and Tuckaleechee Cove.

To the left of the entrance to the Loop Road, ridges unfold in layers or seem to gallop down the hillsides like young colts. The forests are filled with gorgeous color. It is difficult to find a prettier view in the park. In this scenic drive through part of the cove and then over Cades Cove and Rich Mountains into Townsend, there are not only log cabins and split-rail fences, and even open pastures (some seeded with the natural grasses of the area), but also some of the wildlife of the mountains, including white-tailed deer, black bears, skunks, groundhogs, wild turkeys, and numerous other birds. With the splendid foliage lighting up the landscape, this is a place of great peace and beauty. Although the route described in this text will leave the cove before the historic Cable Mill area, the visit to the mill and visitor's center located about halfway around the Loop Road makes an excellent side trip.

Along the Loop Road there are some rounded gray rock exposures beside the road that are outcroppings of the Paleozoic-age limestone that lay beneath the older rock thrust faulted over the limestone. Erosion and removal of the older rocks by weathering has exposed the younger underlying limestone and formed the valley we call Cades Cove (fig. 26).

One of the trip's highlights is along a one-way mountain road crossing Cades Cove Mountain and Rich Mountain, which offers a cross-section of

Fig. 26.
One of the more popular areas of the Smoky Mountains is Cades Cove, where the pioneers and American Indians once settled, hunted, and farmed. This view looking south toward the main crest of the Smoky Mountains is from an overlook along the one-way gravel Rich Mountain Road.

how life was lived here in the past. Rich Mountain, at 3,686 feet, separates Cades Cove from Townsend. Cades Cove Mountain climbs almost as high—to 3,380 feet—and slopes down on the north side to join Rich Mountain. Rich Mountain Road, which crosses Cades Cove Mountain at Cold Spring Gap (2,660 feet), offers extraordinary views of the cove. The road then crosses Rich Mountain at Rich Mountain Gap (elevation 1,960 feet) and continues down into the Townsend area. Another geologic characteristic along this trip is the abundance of karst.

Before Cades Cove was a cove, it was part of a mountain range. As it evolved and descended to what it is today, it became a place for hunting, settlement, and work. Native Americans realized the importance of this sanctuary in the mountains. They lived and hunted here for hundreds, if not thousands, of years before white European settlers exploited the rich opportunities of the area. Natural resources seemed unlimited. It is little wonder that when John Oliver came over the mountains from upper East Tennessee and saw Cades Cove, he decided to set a log cabin in the middle of it and call it home. The Olivers bought this site and settled here in 1826, but the Park Service believes that the cabin, a handhewn log structure, was built in the mid-1850s.

On the periphery of the cove, Oliver found hundreds of mineral springs and mountains stocked with a plentiful supply of wild game. Rivers provided a wealth of large trout and bass. John Oliver had been one of the heroes of the battle of 1812 in New Orleans with Andrew Jackson, and while it may have been the peace and solitude of the area that first drew him, the bountiful land kept him here. Although Oliver, originally from upper East Tennessee's Carter County, was the first to arrive, he was followed by a number of pioneer families. They built field schools and churches, erected gristmills on streams to grind their flour, hunted plentiful game, and had vast supplies of pure water.

Another settler was William "Fighting Billy" Tipton, who understood the vast commercial value of this land. He began buying up Cades Cove land grants from North Carolina land speculators who lived in Carter County. Fighting Billy was no fool. He paid a dollar an acre for the floor of Cades Cove and later sold it back for thousands (Brown, 1990).

The late Randolph Shields, former chairman of the biology department at Maryville College and once the leading authority on Cades Cove's history, always maintained that it is impossible to understand the history of this entire region—and especially Blount County, which contains the borders of the cove—without a thorough knowledge of Cades Cove. That's because so many of the first families coming to the frontier in their search for new ground settled in the cove. Not only did they farm its rich limestone soil, but also they found a solid base for political and economic liaisons. Contrary to popular belief, the first people to settle in the cove were in no way uneducated or inbred. They were both prosperous and well-respected. John Oliver himself was a man of great means, and the Tiptons became one of the most politically powerful families in early East Tennessee history. About 116 families lived in the cove from the eighteenth to the twentieth centuries (Brown, 1990). Their strong work ethic, as well as a religious bent, defined them, according to Shields, as "the real mountain people."

Once, the cove rang with names like Whitehead, Burchfield, and Cable. Old John Cable's mill ground grain for flour and his sash sawmill cut timber. Sheep once roamed the cove, too, creating a profitable market for wool. Some cabins were built there so long ago that they utilized the older building technique of burning the logs into shape—without the use of an ax—instead of hewn logs. Split rails were cut from American Chestnut trees, which were profuse before the turn of the century and up until 1928, when the notorious chestnut blight destroyed them. Old-timers remember the forests looking like "deer's heads," because the stumps of the dead trees were white as antlers.

Some of the language used then has been lost, but a few idioms survive. Furniture and belongings were "plunder"; logs were "wagoned" from the forest; and many cove babies died of "flux." In that time, "taking up the cabin," meant that several strong men would grab up corners of the cabin, lift it, and settle it squarely on its notches.

Sicknesses were usually treated with a powerful potion made of boneset or spicewood herbs. People ate wild mustard greens and "branch lettuce," which was found along "cold branches." Catnip was given to babies to make them sleep.

This trip through Cades Cove and over Cades Cove Mountain and Rich Mountain features traveling through time as well as traveling through space. It is one of the most beautiful of all the road tours because the pioneer spirit seems to survive here so strongly that it can be felt even on the drive along the Loop Road.

Buttercups growing around the base of an old tree, or in a clump in a field, are a sign that once a log cabin stood near that very spot, for buttercups have to be planted by someone first before they can return year after year. Cades Cove's old schoolhouse, the church, and, finally, the graveyard attest to the lives and deaths of people who lived and worked here.

The irregular, mounded topography near the 0.4-mile mark on the Loop Road is believed to be colluvial mudflow debris. Sparks Lane, a connector road that comes in from the left from the other side of the cove, is located at 0.9 miles along the Loop Road.

Just past Sparks Lane, the road comes upon the parking area for the John Oliver Cabin. This is a short walk (about a third of a mile, one way) that features the site of one of the cove log cabins. At the parking lot, looking back toward the Oliver cabin, the tourist is standing under the Great Smoky Mountain Fault plane. The mountains behind Oliver's cabin are alive with color. Yellows bleed into reds, greens, and oranges. Tulip poplars, hickories, red maple, sweetgum, and a variety of other hardwood trees provide the colors. Evergreens add contrast. The mountains are in inviting rows here, ripe with color and history.

Continue along the Loop Road as it winds and rises and falls along the borderland topography. There are numerous outcroppings of the light gray Paleozoic limestone. At mile 1.9 the Loop Road passes by a Park Service maintenance road on the right. This gated road leads to an active karst area of the cove. Among the numerous limestone exposures is the entrance to Gregorys Cave. The cave is gated to protect the cave and its wildlife and habitat. For entry into this cave and any other cave in the Great Smoky Mountains National Park, permission from the park superintendent is required.

Near the 2.5-mile mark on the Loop Road is the Methodist church on the right. Built in 1902, it was constructed in 115 days. Just past the church and along a straight stretch of the Loop Road is Hyatt Lane on the left. This lane is another cross-cove road that connects the two sides of the cove.

The cove has worn down over thousands of years of erosional action. Abrams Creek was probably the ancestral stream here, but what puts the views here into perspective is that the mountains cupping around the cove existed once high over the cove.

At log mile 29.9, turn right onto Rich Mountain Road. The Cades Cove Missionary Baptist Church is on the left, and there is a sign showing where to turn onto this rocky, primitive road. Be careful. This is a one-way road, and it is unimproved. Townsend is twelve miles over this road on the other side of Cades Cove Mountain and Rich Mountain. This trip takes about an hour to drive, and it's imperative to remember that the Park Service locks the gates on either end of the road at dark.

Continuing on around the Loop Road would make for a nice sequel for the trip on another day. Or make the Loop Road the first choice, and reserve

Rich Mountain Road for another time. The rest of this tour describes the route across Cades Cove Mountain and Rich Mountain.

Again, go slowly, because the dirt road is full of curves. Along the route it is possible to see a great deal of wildlife. Even the famous black bear can be seen on this road, but do not get too close. These are wild, incredibly powerful animals. In the spring of 2000, a woman was mauled and killed in the park by two black bears. As the route winds around the switchbacks, the canopy above is a cathedral of color like none other in the United States. There is more biodiversity in the Great Smoky Mountains National Park than anywhere else on the planet, and more varieties of trees here than in all of Europe.

As the sun filters down through the tops of the trees hanging high above, it is possible to imagine that this is a holy place. The soft light turns the leaves to pastel and blurs the colors, and then, just as suddenly, the light shifts through to the greens of the poplar, turning them to an almost neon glow.

As the route climbs higher, it will again be above the fault plane, heading up the southern side of the mountain. Deer here can be so tame and unconcerned about human contact that visitors can climb out of their cars and snap photographs of them.

As the road leads on, note the Rosebay Rhododendron that grows in the shade of the mountains. In the spring, it has a pinkish flower. The Catawba Rhododendron lives on the heath balds in full sun. There is also a great deal of mountain laurel, called mountain ivy by the old-timers, on the south side of the ridges. To distinguish between rhododendron and mountain laurel, the longer name (rhododendron) has a long leaf and the shorter name (laurel) has a short leaf.

One of the real treats of this trip is that from time to time the old roadbed can be glimpsed as it fades in and out of sight. The original roadbed probably began as a game trail, then was adopted by the Indians, and later dominated by the pioneers. The roadbed represents a past that can be encountered at almost every turn in the road.

A beautiful pulloff point for photographs comes at thirty-one miles. It frames such a picturesque view that some camera buffs refer to it as Photographer's Point, because this view looks back down into the valley. Glimmering in the distance, like a white jewel, is the Methodist Church, which is surrounded by color. This is a remarkable view. It is one of the most photographed scenes in the Smokies, a perfect postcard photo.

The next good pulloff is at 33.2 miles, where the far western end of Cades Cove is visible. Just a bit farther on, the route crosses the top of Cades

Cove Mountain, where the slopes down the mountain become very steep. It's hard to concentrate on any one aspect of the view here because the area is immersed in color and the view is overwhelming, almost too much to take in at one time.

The third pulloff is the Foothills Overlook at 33.9 miles. It's on the north side of Cades Cove Mountain as the road winds down to Rich Mountain and on into Townsend. This vantage point is at an angle to look directly into the foothills of the Smokies. At this elevation, the viewpoint is on an equal level with the mountains wavering in the distance. It's an exhilarating moment.

One other exciting feature here is that from this view, it is possible to see the Valley and Ridge Province behind the Blue Ridge Province. And on a very clear day, even the Cumberland Plateau appears in the distance, jutting out on the horizon. This area is virtually saturated in fall color. Note the broadleaf hickories that look as if great chunks of leather or tobacco leaves hang from their limbs. Also notice the masses of green ferns growing wild on the banks near the roadside. At times the falling leaves simply cover the road in yellow and gold, like the swipe of a paintbrush.

At this point in the trip, the route passes through a sinkhole that covers roughly five hundred acres. Think of being in a hole deep enough to swallow an entire city block. That's how wide and deep this sinkhole is. In fact, Bull Cave (located at 36.7 miles) is so deep, you have to drop hundreds of feet down into a funnel-shaped depression just to get into it. Bull Cave is formed in the fractured rock of the Knox Group. Most of the fracturing was caused by the action of the Great Smoky Fault grinding along the rock surface. The cave, located in the park and protected by park law, is believed to be one of the deepest caves in the eastern United States, plunging almost 700 feet down and over 4,300 feet in length.

It is a vertical cave requiring considerable caving experience to explore. Mountaineering techniques such as rappelling and ascending methods are required in this cave, which also has a waterfall. Again, any visitor must have a permit from the park superintendent to enter any park cave.

At the park boundary, Rich Mountain Road passes through Rich Mountain Gap just at the parking lot for Bull Cave. At the gap, note the old block cabin on the right. It once belonged to the late Tillman Cadle, a famous coal miner who was in every major coalfield battle of the 1920s and 1930s. He was a United Mine Workers of America organizer and a dead shot with either pistol or rifle. He was also famous for traveling with his wife, Mary Elizabeth Barnicle, throughout the Appalachians in the 1920s, collecting songs, ballads, and church sermons. This was before the Library of

Fig. 27.
Rich Mountain provides a backdrop for the monuments in this cemetery in Townsend.

Congress and the American Council of Learned Societies sent John Lomax and later his son Alan into the South and southern mountains collecting blues and spirituals, work chants, folk tales, and ballads (Ferris, 1989).

As the road winds down the side of Rich Mountain there are many outcrops of the light gray limestone that is characteristic of the Knox Group.

Do not turn right where Dry Valley Road connects to Rich Mountain Road. Stay straight here. Then at 40.2 miles, turn left off of Rich Mountain Road, where there are some horse stables on the right.

At 41.0 miles, turn right. The Tuckaleechee Methodist Church Cemetery is on the right at the turn onto Old Tuckaleechee Road (fig. 27). This will lead to State Highway 73. Turn left onto State Highway 73 for the return to Maryville. Campground United Methodist Church is on the right at the intersection with S.R. 73.

The natural history of this trip is so fascinating, it calls out for visitors to take their time, get out of their cars, and examine some of the rocks and outcroppings and see the panoramic sweeps of color. This is the best way to step back in time, rub a hand across an ancient ocean floor, and feel the roll and thunder of billions of years coming together in one trip.

PART II

The Valley and Ridge

Elrod Falls

Destination: *Elrod Falls in Hancock County.*

Route: *Beginning and ending in downtown Knoxville—I-40 and I-81, U.S. 25E, U.S. 11W, S.R. 66, and S.R. 31 (includes Hancock, Hawkins, and Grainger Counties).*

Cities and towns: *Bean Station, Mooresburg, Rogersville, Sneedville, Rutledge, and Blaine.*

Trip length: *Approximately 185 miles roundtrip.*

Nature of the roads: *Four-lane interstate and state highways, two-lane paved state highways, very curvy in Hawkins and Hancock Counties; short length of gravel road approaching Elrod Falls (about 1.4 miles).*

Special features: *Elrod Falls, scenic Clinch River, historic Rogersville, Rutledge, Bean Station, and Sneedville; pastoral scenes along S.R. 66 reflecting Appalachian heritage.*

Fall is a time of shorter days and longer nights. The sun is heading south again, and the weather turns cooler. Deciduous forests quickly begin to change in color as light and temperatures drop. It is the forest's way of announcing the arrival of fall.

The process itself is neither simple nor quick. The green leaves of summer, full of chlorophyll and the yellow pigment carotene and xanthophyll, begin to change gradually with the chilly nights. Chlorophyll production slows, breaks down, and green departs as yellow pigment begins to shine through the leaves.

For trees like blackgum and maple, cooler nights begin the conversion of sugar and other compounds that change their leaves to orange, red, and maroon, depending on the mixture of red and yellow pigments. As the bard said, "And this our life exempt from public haunt/Finds tongues in trees, books in the running brooks/Sermons in stones, and good in everything" (*As You Like It*, II.1).

Now, with the sense of urgency fall brings, is the time to find sermons in stones and listen to the trees before winter sets in. In these last crisp days of autumn, explore the temporary burst of beauty that the waning season offers. And one of the best places to find it is on a trip to Elrod Falls in Hancock County up along the shoulder joints of Clinch Mountain where the scenery will supply plenty of color for warm memories in the cold months.

But don't go asking Bud Brummet, a resident of Hancock County, for directions. He enjoys the peace and quiet of the hills above Sneedville, where he grew up. He likes nothing better than to watch his mule, horse, and dog play while this year's tobacco crop dries on locust poles in the fields rather than in a tobacco barn. In today's economy, it is cheaper to cure tobacco in the open than in a barn in this part of East Tennessee, because barns are too expensive to build these days, and hiring all the hands it takes to put up tobacco is becoming too costly as well. So hanging the shaggy leaves on locust poles in a field is cheaper, though it signals the end of an old, traditional ritual in the hills of Tennessee.

From Knoxville the drive to Elrod Falls takes a little over two hours. On the return trip, any of several directions lead back to Knoxville, making the trip an all-day affair. Make a day of it.

The itinerary takes you across several ridges and two mountains, all of which are located in the Valley and Ridge Province of East Tennessee. Where we begin in Knoxville, the road is a six-lane interstate that soon narrows into a four-lane highway and finally into a skinny two-lane ribbon that twists over ridges and across valleys.

For those making the trip from Upper East Tennessee, there are two routes. Either come down I-81 to the intersection of U.S. 25E and proceed, or come down from Kingsport on U.S. 11W and join the trip at State Route 66 along the Rogersville Bypass.

Rocks of varying types and ages can be seen along the route. Limestone, shale, and sandstone underlie the ridges and valleys and form such high places as Stone Mountain, Clinch Mountain, Copper Ridge, War Ridge, and Chestnut Ridge. The valleys are floored with shale and limestone.

From Knoxville, take Interstate 40 to I-81 (map 6) and go north on I-81. Exit onto U.S. 25E, about fifty-six miles from Knoxville. This is the exit to Walters State Community College. Go north toward Morristown, where rolling hills and some color in the long, pastured views become visible. This topography is rolling because of the limestone and shale that underlies the landscape, where erosion resulted in the pronounced ridges and valleys. The route from Knoxville runs parallel to the general strike of the bedrock.

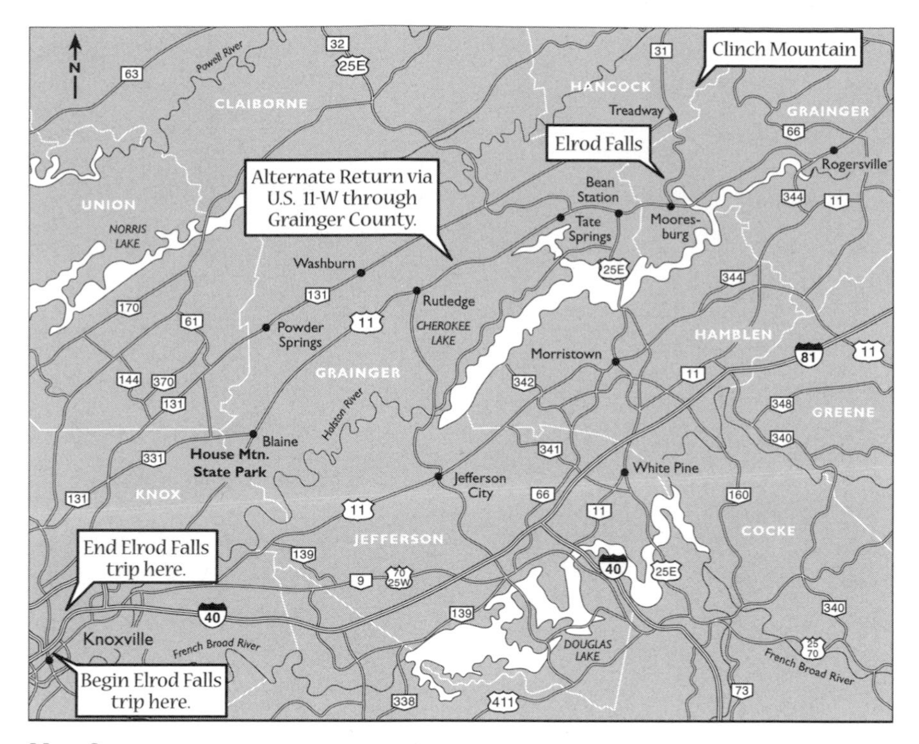

Map 6.
The Elrod Falls trip.

Once the route leaves the Interstate, it begins to cross the strike of the rock layers, rising over the resistant rocks, like limestone and sandstone, and dipping down over the shale areas. Crossing old highway 160, the route passes over a series of valleys and ridges, a pattern that holds for the entire trip.

About sixty-one miles from Knoxville, Clinch Mountain seems to loom suddenly before your eyes. The color here is brilliant. On the mountain tulip poplars and buckeyes have already turned yellow.

Another two miles and the route crosses the Holston River and leads into Grainger County, where, at the Hamblen and Grainger County line, numerous layers of grayish white to medium-gray rocks are evident along the banks of Cherokee Lake. These rocks are composed of Cambrian- and Ordovician-age limestone, formed in a shallow sea some 400 to 500 million years ago. Take the time to absorb the beautiful scenery outlined against Cherokee Lake and the Holston River.

After stopping at Cherokee Lake, drive up over the ridge and drop down into Bean Station. Just below the smoky wreath of clouds hanging on the tips of the mountains, the descent opens up a vista of color along the knobs.

At Bean Station, turn right onto the exit for Rogersville on U.S. 11W north. The route heads toward Mooresburg before leaving Bean Station. Short Mountain is on the horizon. It's the one with the whitish scar near the crest, the site of a silica mine, where quartz-rich sandstone of the Clinch Formation is being mined to make glass. Sandstone forms the bulk of Clinch Mountain, dividing portions of Grainger, Hancock, and Hawkins Counties and providing the resistant "back" for Short Mountain.

Mooresburg is still in the valley. On the left is Short Mountain and on the right is Prophet Ridge, just on the edge of Hawkins County. At roughly seventy-six miles from Knoxville, the route crosses Poor Valley Creek, and Short Mountain now appears on the left.

Here the trip enters sinkhole—or karst—country, an area about a mile wide and five miles long. Sinkholes are all around, on both sides of the highway. Actually, the deepest and widest sinkholes lie beneath the roadbed.

Sinkholes are the result of chemical weathering of limestone when acidic groundwater dissolves the limestone along cracks in the rock. Over time, perhaps hundreds of thousands of years, the cracks enlarge and begin to interconnect, forming caves. As surface soil is eroded, it is also carried down into the cavity-laden limestone, resulting in the formation of surface depressions and sinkholes.

On the left now is Stone Mountain. At about eighty-two miles, when the route crosses Clouds Creek, Town Knobs appears immediately in the foreground. The color here is always amazing, even in years when fall color is poor. The Holston River on the right offers striking contrasts of blue water and the blazing reflection of autumn's trees.

Next you will come into Rogersville, which is definitely worth taking time to explore as a side trip. Rogersville is one of East Tennessee's most historic towns, boasting many old and beautiful homes. The Hale Springs Inn should not be missed. Among its guests have been Presidents Andrew Jackson and Andrew Johnson.

After touring Rogersville, turn right onto State Highway 66 at roughly eighty-eight miles from Knoxville. Be careful to follow the exit ramp from U.S. 11W (along the Rogersville bypass) on around and then turn right, or north, onto highway 66. The highway is just a big loop to the right. This route will pass back over highway 11W, at which point it enters the switchbacks of Town Knobs and Stone Mountain, the same knobs seen earlier from a distance. The highway is virtually covered with a canopy of trees. But be careful here, because highway 66 is a two-lane road with numerous curves. The original roadbed was likely a wagon road carved into the contours of the land.

The hillsides in Townsend display a palette of mixed colors during the fall season. U.S. 321 is shown in the foreground.

Rich Mountain Road offers stunning views of leaf color as it crosses the mountain out of Cades Cove.

Some maple leaves turn a bright yellow and contrast perfectly with the fall blue sky.

Above: A scenic overlook of the foothills section of the Smoky Mountains is found along the Rich Mountain Road as it descends into the Townsend area.

Late fall offers opportunities to view the final stages in leaf color change when once vibrant colors turn to a buff color, as these once yellow maple leaves show.

Right: Beech trees display a vibrant yellow and burnt orange color during the fall season.

Below: Mountain roads in East Tennessee, such as U.S.23/I-26 in Unicoi County, offer the best opportunity to see the turning of the leaves during fall.

The mixed deciduous forest, especially along the rivers, such as the Nolichucky River in Unicoi and Washington Counties, can present some of the best color for photographs.

Late summer and fall can bring fog to some of the valleys of East Tennessee and a hint of the coming winter season. View from the top of Clinch Mountain in Grainger County.

Redbud leaves offer a platform for the morning dew in fall.

A stream-worn boulder serves as a canvas for this spectrum of fall color.

Above: Abrams Falls is framed by golden yellow hues of early autumn.

Right: These leaves seem to float in air as their color blends into the underlying stream currents.

After leaving the thin-layered shale and sandstone hills of Town Knobs, glance over the valley to Stone Mountain, where sandstone beds build the mountain slopes.

On this leg of the trip, there are many sights of rural Tennessee to enjoy: farms, fields of rocks, corn browning in rows in the fields, and neat, golden rolls of hay. Many of the barns in this area rest on rock foundations. Tobacco baskets hang on the sides of bleached barn boards, and in the fall the goldenrod is in bloom, creating a pleasing blend of browns and golds. Note the grayback limestone jutting up in the fields.

Soon the route enters Cedar Valley, a road lined with mixed hardwoods of walnut, sycamore, maple, and dogwood. Old cedars (junipers) sprout here and there. The shaley limestone and shale strata that are found here are often fossiliferous, containing the remains of brachiopod shells and trilobites (fig. 28).

While crossing Stone Mountain, the route curves its way to the top and then down again, into Poor Valley. Most of the color on Stone Mountain is provided by tulip poplars and maples.

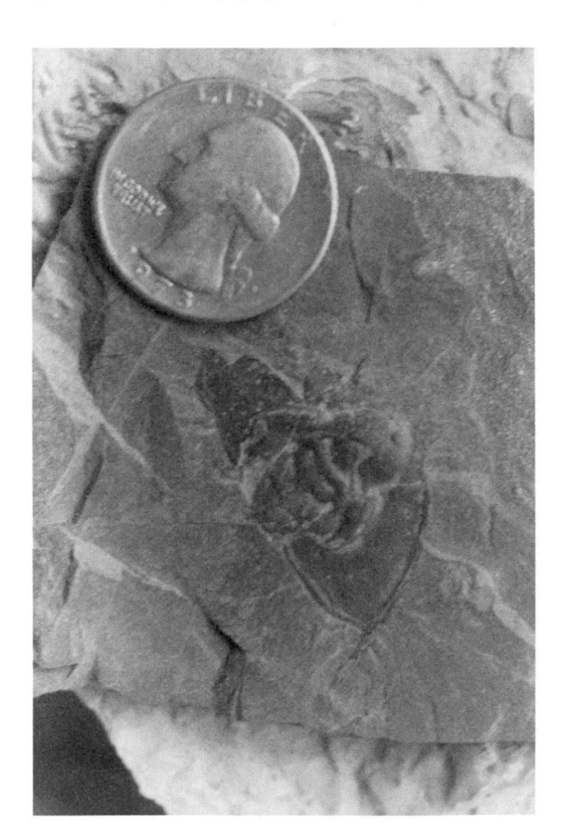

Fig. 28.
Fossil trilobites can be found in the grayish black shale of the Rome Formation exposed in Hancock County. This is the head portion (cephalon) of the trilobite Olenellus.

Poor Valley gets its name—so the story goes—because the soil is too poor for farming. One reason is that the land sits on top of Chattanooga Shale, which also contains oil and uranium deposits. Consequently, the soil is thin, and soil mixed with shale is suitable for nothing but weeds. But in Richland Valley (the broad valley in Grainger County just south of Clinch Mountain and Poor Valley Knobs), land sits on top of limestone, which enriches soil and makes it fertile.

At 103 miles from Knoxville, this tour comes to the crest of Clinch Mountain, where, in autumn, the roadside is a blue blur of chicory. This is the same Clinch Mountain that rises up in Grainger County to the southwest and is crossed by U.S. 25E. Numerous layers of white to light brown and gray sandstone are visible on the carved mountainside. Those layers disappear once the route crosses the crest of the ridge and starts down the other side. Thin-bedded shale and limestone, exposed in the cut slopes along the winding road, dominate the north side of the mountain. Figure 29 illustrates the geologic structure of the Hawkins and Hancock County area along State Highway 66.

And Copper Ridge, the next area the route comes to, is built up of Knox Group limestone and dolostone. The road follows the creek, which over centuries has carved a water gap through the ridge.

Rocks found here were formed in a shallow sea environment where ocean waters pounded the shores in rhythmic motion and algae grew in the tidal areas. The algae formed layered "mats" of sediment, which have been transformed into hard silica-rich forms in the rock called stromatolites (also called cryptozoon). These masses often form chert deposits in the weathered limestone and dolostone and can be seen as whitish gray rocks in the reddish orange clay soils.

Here, along the road, the colors of human-made equipment blend into the fading landscape. Brown barns with rusted tin roofs lean out near the road and rusting mule-drawn rakes and threshers, relics of another era, sit in fields like oxidized skeletons. Sumac, which has already reddened, brightens the roadside and the rocky hillside. Cutting through the sandstone gaps, yellow poplar leaves display themselves.

Pass Richardson's Creek Baptist Church on the edge of Lee Valley. The church and its picturesque setting deserve inspection. A local resident, Bill Jennings, who made firebrick in Missouri until he got tired of the winters and retired to Lee Valley a few years ago, looks out from his barn to greet visitors. "You couldn't dynamite me out of here now," he says, thumbs hooked under the straps of his overalls. "I got 12 acres of land and I don't do nothing, but just what I have to do around here."

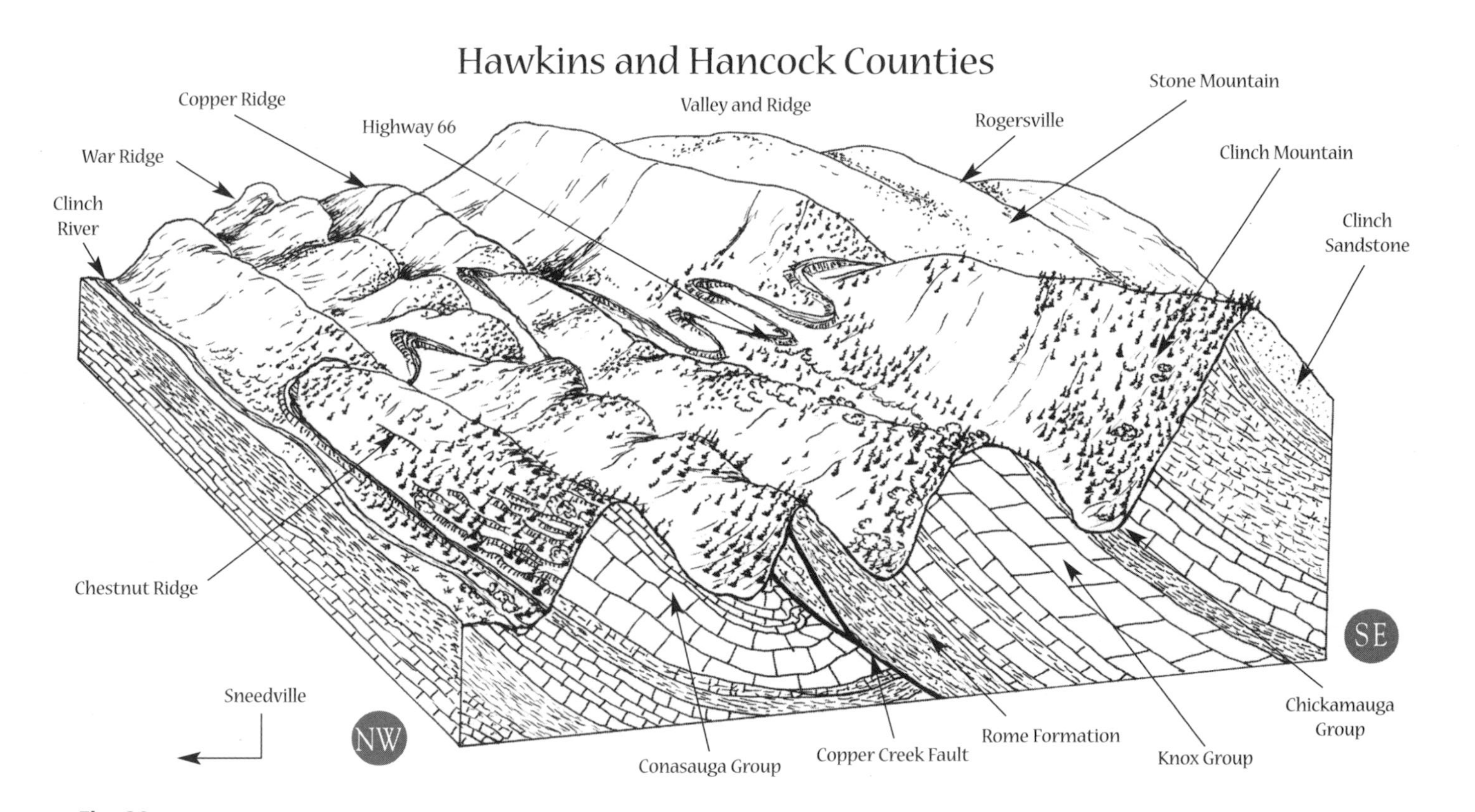

Fig. 29.
The underlying geologic structure of the landscape along State Highway 66 in Hawkins and Hancock counties is shown on this generalized schematic diagram.

Somewhere around 105 miles from Knoxville, the trip leads to War Ridge and then drops down into the next valley. Here again the valley floor is underlain with shale and shaley limestone, which settlers tilled for their livelihood.

In the next mile, the route enters Hancock County, where the roads are thick with hardwoods on both sides. Mountain streams in this area drain northward into the mighty Clinch River. The route then passes through Trent Valley and across Chestnut Ridge, a place of almost perfect isolation.

This natural isolation made the settlers extremely self-sufficient. Steep ridges presented barriers for people, and as a result the people tended to stay in the valleys, traveling up and down the valleys instead of across the ridges. Today, although the roads generally follow the valleys, every now and then the roads will cross the ridges, which still present a barrier for construction.

At roughly 111 miles from Knoxville, the route begins to run parallel to the Clinch River and its floodplain. Parts of highway 31 flood at times, because there is no dam on the Clinch River north of the new bridge at Sneedville.

The rivers and their valleys were key to the settlement of Hancock County. In 1795, when the first white settlers began to flow in a steady stream from Virginia, North Carolina, Kentucky, and South Carolina, they located their log cabins near water, which originally cut through the valleys and the mountains.

In this area, the Clinch River cuts across the entire county from northeast to southwest, and the Powell River nips the northwest corner of the county. Even though settlement began early, the county has remained sparsely populated. Sneedville is still the only town in the county, and for years its mayor harangued the Tennessee State Legislature to "either build me a road in or a road out."

Many of the first pioneers, of whom there is little record, settled along Newman's Ridge, whose inhabitants have puzzled genealogists for hundreds of years. These people are known as Melungeons, identified by their straight black hair and coppery, olive-colored skin (Price, 1971). By blood, they are neither white, Indian, nor African American. Their eyes are dark black, brown, purple, or gray. Their high cheekbones indicate Indian ancestry, but they have neither round faces, nor broad foreheads, but thin lips and narrow faces.

The Melungeon men were almost always thin, and the women were described as strikingly beautiful. Over the years, white settlers were attracted to the strange, alluring beauty of the Melungeon women, married them, and, for the next two hundred years, watered down the bloodlines. Today, no pure Melungeon people are left in Hancock County, and little is known about their history.

Sneedville was called Greasy Rock by pioneers and long hunters migrating to this rich wilderness. The town earned its name when pathfinders shot game and cleaned it on a large rock at a spring just below the town.

At 112 miles from Knoxville, turn right to cross the Clinch River over a new state bridge into Sneedville. One memorable stop is at Aunt B's for lunch. The restaurant is in downtown Sneedville. You will miss good food, but mostly entertainment, if you miss Aunt B. Her wit is worth the price of any meal. Bring your best stories along with your appetite. Aunt B's is a sanctuary for the hungry of spirit and stomach. Home folks will be sitting around tables hunkered down over platters of country cooked, fresh vegetables and potatoes that are peeled and not poured-out flakes from a box. Here the hungry traveler can find homemade cornbread, pies and cakes, and sweet tea that tastes like tea and not industrial runoff. But the tea comes in only one variety—sweet.

After leaving Sneedville, take highway 31 back toward Knoxville. Highway 31 leads back up Chestnut Ridge. War Ridge will be on the left. This is the object of the entire trip, and, unfortunately, the way there is not clearly marked. The sign to Elrod Falls is not easily seen from the roadside.

A good way to judge the distance to the turnoff for the falls is to clock the mileage from War Creek Baptist Church, on highway 31, which is about 8 miles from the bridge. The turnoff to the falls is about 1.2 miles from the church, going toward Mooresburg and Bean Station.

At roughly 121 miles from the beginning of the trip, turn right from the new highway onto the old highway (about 1.2 miles from the church). Here, go left and drive about a quarter mile to a gravel road on the right. Look for a small wooden sign with orange lettering on the right side of the road. It reads "Falls 1.2 miles." Turn right here and go about seven-tenths of a mile on the gravel road, which winds through farm country. Soon the gravel road comes to another very small sign that points to the left. The drive down that second gravel road features ironweed and thistle growing along the roadside.

This road takes you to a small parking lot that is about 100 yards from Elrod Falls, and the roar of the falls can be heard from the parking lot. All of this area comprises a park owned by Hancock County. Although it is in rather shabby condition at the moment, there is an open shed for picnics. This site is at the base of Copper Ridge.

To reach the falls, simply follow the small path along the creek up the hollow. This is a footpath, about one hundred yards long, which gently climbs to the base of the falls. This is the highlight of the entire trip. Hardwood trees of every variety surround the cascading waterfall that plunges over giant ledges of limestone bedrock. These are the same kinds of limestone and dolostone layers of rock that underlie Copper Ridge, which the route crossed earlier in the trip.

Lobelia, jewelweed, and purple asters flourish just below the falls, and ferns abound in the shady area around the dam.

According to those familiar with the park's history, this land once belonged to a farmer named Brown Johns, who sold it to Hancock County for about eight thousand dollars.

"There are actually three falls here," says Teresa Yount, who is working to save Elrod Falls from neglect and abuse. "The second falls is about fifty feet higher up. You can actually walk beneath it." A third falls is straight up Copper Ridge and difficult to reach.

After leaving Elrod Falls (by retracing the route back to the main highway and then turning right on State Highway 31), the route crests Copper Ridge, the same ridge that runs down through Knoxville. Cross Flat Gap on highway 31 and then head up the scarp side of Clinch Mountain, crossing into Hawkins County, where there is a stunning view of English Mountain and the Great Smoky Mountains.

From this viewpoint Mount LeConte can be seen wavering in the distant blue haze. The trip now leads into East Poor Valley and then back to Mooresburg and 11W. At Mooresburg, turn right on U.S. 11W and stay on 11W south and follow it into Grainger County. (Do not turn onto U.S. 25E toward Morristown.) On the right 11W is parallel to Clinch Mountain's lower side. On the left is Big Ridge, with Cherokee Lake gleaming like a jewel in the distance.

Pass Kingwood School, with its wonderful gazebo, on the right. In this spot, the old Tate Springs Hotel once flourished. People from across the nation came to the hotel to drink its water, much of which was filtered through Chattanooga Shale, which loaded the water with iron, minerals, and, unbeknownst to the old-timers, oil and uranium.

While passing through the town of Rutledge in Grainger County, you might want to stop and buy a basket of luscious Grainger County tomatoes, which are as famous to the area as sweet onions are to Vidalia, Georgia. A side trip to the town's old jail makes an interesting visit. Built in 1848, the building is the oldest solid brick jail standing in the state. Although the holding cells are gone, the original door and windows remain.

The Elrod Falls trip is a treat for photographers, historians, and amateur geologists. The hundreds of formations throughout this ridge and valley excursion supply endless possibilities for study, and the history of the region yields fascinating stories. It's a trip that will remain in memory long after it's over.

Clinch Mountain

Destination: *Crest of Clinch Mountain in Grainger County.*

Route: *Beginning in downtown Knoxville–U.S. 441, S.R. 131, U.S. 25E, U.S. 11W (Knox, Union, and Grainger Counties).*

Cities and towns: *Luttrell, Powder Springs, Washburn, Thorn Hill, Bean Station, Rutledge, and Blaine.*

Trip length: *Approximately ninety-seven miles.*

Nature of the roads: *Four-lane city streets, four-lane rural highway, two-lane paved state highways; some curvy sections, no paved shoulders on rural areas of S.R. 131.*

Special features: *Clinch Mountain scenic overlook, rural countryside.*

Listen to the poetic geologic language associated with the Valley and Ridge Province: Water gap, wind gap, axis of the valley, Paleozoic, Cambrian, Mississippian, Silurian age, strike of the strata, dip angle, moss animal, *Arthrophycus*. This is a trip that transports the tourist across millions of years of the earth's history by viewing the present. At the summit of Clinch Mountain is coastal marine sandstone that is 430 million years old. At the peak, the visitor is actually on top of the bottom, and if it were the Paleozoic age, this spot would be under water, on an ancient sea floor that is only one-eighth as old as the Earth. This is the Valley and Ridge, a land rich in graceful language and in geologic patterns.

This tour is replete with languid hillside farms. Rolled bales of hay dot the greensward like brown, shaggy beasts. A short distance on the other side of Luttrell, log barns pop up like chocolate drops. They are stuffed with hanging stalks of tobacco, curing in the sweet autumn air.

This road trip cuts across beautiful valleys and climbs over a large mountain that lies just northeast of Knoxville. The route is completely within the Valley and Ridge Province and consists of roads that principally follow the axis of the valleys while having to cross the ridges and mountains at water gaps or wind gaps. Upturned layers of Paleozoic-age sedimentary rocks generally underlie the landscape. This means the layers of rock have been displaced

from their original horizontal position and turned up so that the layers are at an angle with the surface of the ground. This violent action took place with the collision of the continents at the close of the Paleozoic Era. The rocks vary from sandstone, siltstone, and shale to limestone and dolostone. Dolostone is magnesium calcium carbonate containing the mineral dolomite and limestone is a calcium carbonate having the mineral calcite.

First up is Luttrell to Thorn Hill and across Clinch Mountain. This is one of those geologically fascinating trips, for it spans Paleozoic time, from 550 million years to 330 million years ago, within a single six-mile stretch. As a whole, the trip goes from about 530 million years to 300 million years before the present, covering about 250 million years, changing from land to deep sea to shallow sea. Geologists from all over the world study the Thorn Hill section, because this amazing change in time in such a short distance is sequential. What was there was preserved. There are just not a lot of places on earth where such a span of time can be seen in such a short distance.

To start this tour from Knoxville follow State Highway 33 to Halls Crossroads (map 7). In Halls turn right onto State Highway 131 (Emory Road) and follow it to the intersection with State Highway 331 at Harbisons Crossroads. Stay on S.R. 131 and then turn left at Harbisons Crossroads and go toward Luttrell.

The route so far has traveled across gently rolling terrain underlain by folded and faulted limestone and shale strata, mostly of Cambrian and Ordovician age. These rocks tend to be worn down over time to low, rounded knobs and broad, spreading valleys. Thick sprouts of bedrock, looking like solid toadstools, are not common but are likely to be seen along the road and in the adjacent pastures and near the woods line.

As you approach the historic Gibbs community—first settled in 1792 by Nicholas Gibbs, a Revolutionary War veteran—near the high school, the road crosses very close to Campbell Cave. The cave was developed in Chickamauga Formation limestone, a rock that was created in a shallow sea some 450 million years ago. The cave has several entrances and is about 300 feet in length.

Not far past the Gibbs community highway 131 joins State Highway 61 in Luttrell. Turn left in Luttrell and stay on highway 131. Just outside of Luttrell, highway 61 turns left to go to Maynardville; however, stay on highway 131 and drive up the valley through Powder Springs and Washburn to the Thorn Hill Community.

In Luttrell, notice off to the right (heading toward Washburn) the tall steel crane towers, anchored securely by cables. The towers are located in an

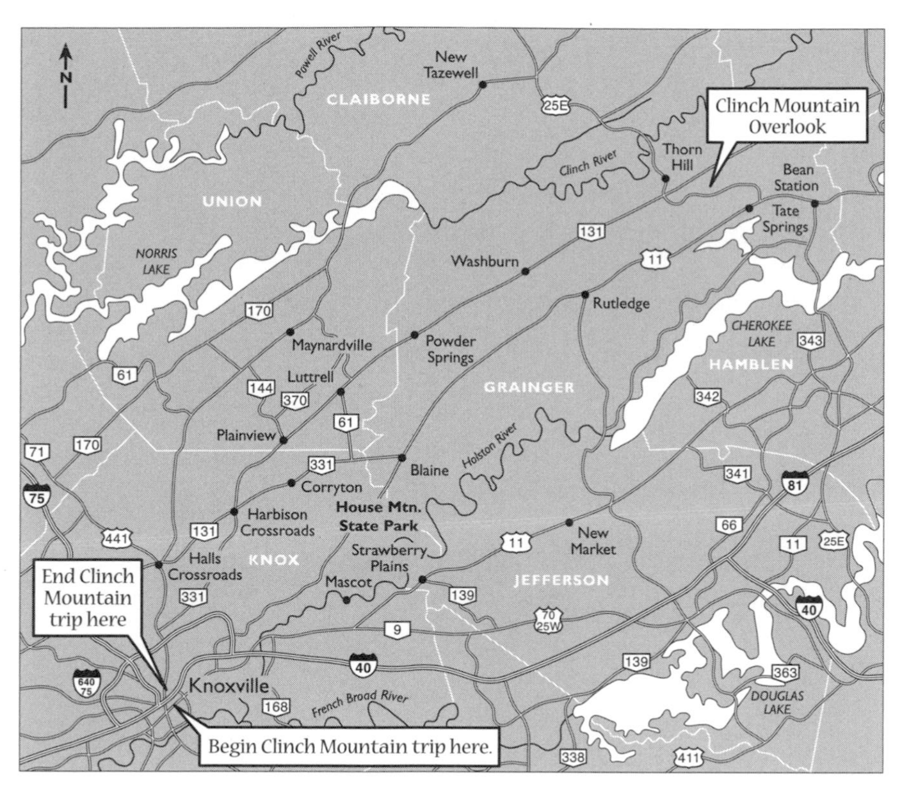

Map 7.
The Clinch Mountain trip.

old marble quarry where Tennessee marble was once mined. This pink to pinkish gray rock, often sold in unrefined form, is a coarse crystalline limestone (the crystals being spary calcite–a variety of the mineral calcite). Once polished, Tennessee marble, prized for its soft pink beauty, graces many of the nation's buildings in Washington. The limestone polishes like glass. Its thick, bedded character also lends itself to being saw-slabed, making it a very decorative building stone. The limestone has been used for buildings all across the United States, including the hallway walls in the Montana state capitol building and the Smithsonian Museum of Art in Washington, D.C.

The quarried pinkish limestone has a colorful past. It was once an ocean reef, called a Bryozoan reef, which is more than 400 million years old. The reef was part of a complex of mounds formed by an invertebrate animal known as moss animals. The moss animals were around at a time when life had not evolved to the point that there were animals crowded over the land yet. But soon thereafter animals did begin to crawl out of the water.

Continuing on up the valley floor notice the numerous grayish limestone outcrops, solid humps in the ground looking like speckled gnomes in the passing fields. The ground is so rocky and the soil so shallow that only the tough cedar tree can gain a grip and grow with any degree of success. When the first settlers here cleared the land of trees, the land was more rock than pasture. What these settlers plowed along present-day highway 131 was once a shallow sea in the early Paleozoic.

As the route crosses from Union County into Grainger County it passes the home of Grace Phipps. Her home and land form a kind of stitch across the county boundary. For more than three decades, she and her husband, John, farmed this land between the Southern Railroad line and highway 131. Three giant sugar maple trees guard the roadside near what was once her home. They are always in full color in fall. She said the trees reminded her of John, who planted them when they first moved to the land.

"Fall is the saddest time for me," she once said. "John loved it, though. He loved his roses and the fall. We built this house in the fall and our children were born here." John was seventy-four years old when he died during one fall season. And Grace never remarried.

Highway 131 follows Clinch Valley past Thorn Hill and into Hancock County. The highway is squeezed in between Clinch Mountain on the right and Copper Ridge on the left. Washburn and Puncheon Camp are two more communities established along this roadway.

Puncheon Camp Baptist Church is located in the Puncheon Camp Community, just up the valley from Washburn. This is one of the oldest churches in Grainger County; it was established in 1804. *The History of the Churches of Grainger County* (Moore, 1986) notes that "[t]he first church building was made of logs and hewn out of puncheons for the floor. . . ." In 1880 Dock Roberts, a local community resident, donated additional land to place the existing structure upon. He signed a deed that made it plain that if it ceased to be a church it was to go back to Roberts's heirs.

The landscape had a great effect on the people of the area. The tall, linear Clinch Mountain provided a natural boundary between the residents of this area. Access over the mountain was formidable at best, and very few people regularly traveled over the rock ridge in early times. These natural barriers even persisted on the north side of the mountain as Copper Ridge, War Ridge, Log Mountain, and Chestnut Ridge. The settlers lived in the valleys and farmed the fertile soil of the valley floor.

Because of the ridges, the people developed a dependency on one another in order to survive the harsh conditions during winter. Neighbors became very close and tended to shut out the rest of the world. When someone unknown wandered into the valley everyone became skeptical of the "intruder." Being suspicious of visitors was a common trait back in the early days, but once the visitor was known and trusted then a bond of iron would join the people and could not even be broken at death. Traveling along this route it is easy to see how the geologic character of the landscape helped to form the culture of this Appalachian community.

Past Puncheon Camp, Clinch Mountain begins to rise off in the distance, running like a great wall of color. Stay on 131 until it reaches Thorn Hill. Turn right onto U.S. 25E.

After the right turn on 25E at Thorn Hill, the route goes up the side of Clinch Mountain. Keep an eye out for the Clinch Mountain Restaurant at the top of the mountain. Here is a scenic overlook into a valley that was a warpath for the Cherokees and an area that Boone and Crockett traveled through.

Grainger County, unlike Hawkins, is a fertile land, and Richland Valley is an apt name for a section that can grow just about anything, especially tomatoes. Farms, settled long ago, look like family farms are supposed to look: they feature old houses nestled in gently rolling, green folds, captured by the hills in scenes that seem almost painted. The parallel ridges and narrow valleys have always been abundant in life and growth, and they still are.

The fall color route to this point has generally paralleled the strike of the bedrock across Knox, Union, and Grainger Counties. In other words, the trip has followed the course or bearing of the outcrop of an inclined bed of rock on the ground surface. The first settlers followed the valley floors, which are also parallel to the outcrop of the rock strata. The rock, being inclined, dips back to the southeast, which would make the rock dive under Clinch Mountain. The rocks exposed at the surface trend parallel to the valley and ridges. This is the strike of the strata.

Now in the Thorn Hill Community, the route will turn perpendicular to the strike of the rock and go toward Clinch Mountain (fig. 30). The group of rocks that are found from the bottom of the valley (at Indian Creek, about two miles north on highway 25E) to Clinch Mountain and then over the mountain down to near highway 11W is referred to by geologists as the Thorn Hill section of bedrock.

This section of bedrock contains a continuous accumulation of strata that spans from the Cambrian period (550 million years ago) to the Mississippian period (approximately 300 million years ago). As the route heads south on highway 25E, the bedrock becomes younger in age until it reaches the Saltville Fault (near highway 11W). At this point the Saltville Fault has

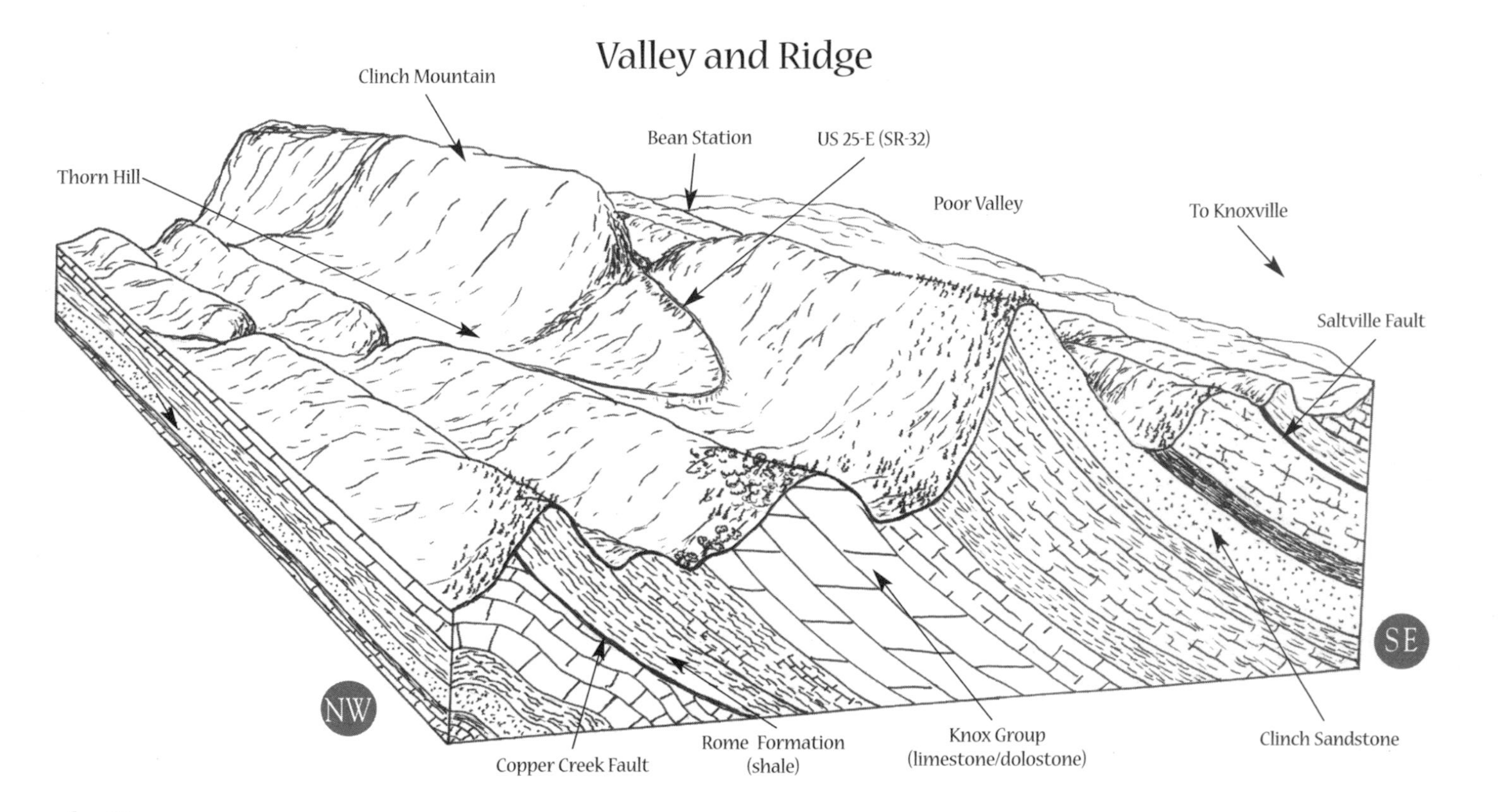

Fig. 30.
The geologic structure underlying Clinch Mountain is shown in this schematic block diagram. The range of geologic formations from Indian Creek to U.S. 11W across Clinch Mountain is referred to as the Thorn Hill Section by geologists.

thrust Cambrian-age strata (the Rome Formation) over Mississippian-age strata (the Grainger Formation). The Cambrian-age strata are over the younger Mississippian-age strata; the fault is the contact between the two kinds and ages of rock.

The ascent of Clinch Mountain gradually brings the valley and ridge landscape characteristic of this area into sharp focus. Because of the mix of deciduous trees and evergreen trees, the fall color display in this area turns the landscape into something resembling a pastel painting.

After crossing the crest of the mountain take the exit to the scenic overlook to the right. Pull off the highway into the scenic overlook parking area.

The view looks out toward Cherokee Lake, which opens below like a great bird on the wing. On clear days the Great Smoky Mountains can be seen from this vista, and the area where Bean Fort once stood can also be observed. In front of the fort was Bean Station and Tavern, the largest roadside inn between Washington and New Orleans in the nineteenth century.

The ridges rolling out in ripples are reflected in Cherokee Lake as it wanders about in angles, twists, and turns. German Creek Bridge and Dollar Island, Bays Mountain, and beyond are all visible.

Amazingly, on a clear day it is possible to see across the Valley and Ridge Province to the Blue Ridge Province, thirty-five miles away. The ridges and valleys are running parallel, a situation created by erosion of the folded sedimentary rock. The German Creek embayment of Cherokee Lake is easily seen from the overlook.

The rock wall along the outside edge of the overlook is faced with sandstone from the Clinch Formation. The unusual feature of this wall is that the sandstone contains the fossil of *Arthrophycus* (fig. 31), which is thought to

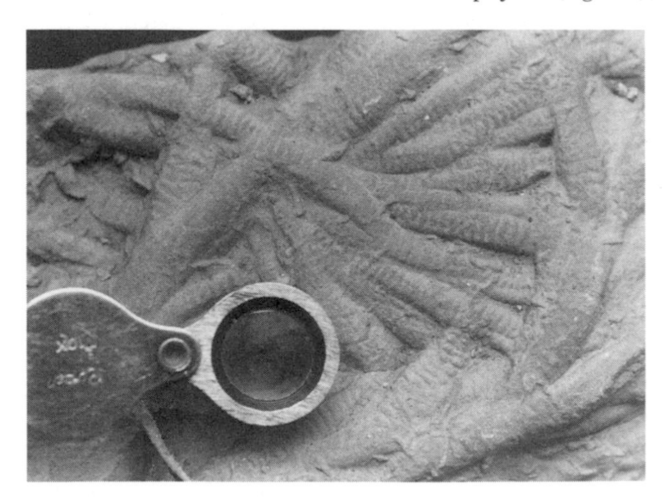

Fig. 31. *The retaining wall at the overlook at the top of Clinch Mountain on U.S. Highway 25E is constructed of sandstone slabs of rock from the Clinch Sandstone and contains the Silurian age fossil Arthrophycus.*

be the burrow of a marine worm or arthropod, or possibly some other burrowing animal.

The animal burrowed in the near shore sand deposit and horizontally mined the sand for nutrients. As the animal moved back and forth through the sand, it left behind a deposit by moving its appendages so that the burrow was packed by the "mined" sediments. This produced a peculiar pattern in the sediments that look like "ribbed" worms crawled over each other in a fight for the best place to be.

Arthrophycus is an index fossil for the Silurian age (approximately 425 million years old). Silurian, a time when the first plants appeared on earth, is the Roman name for a Welsh tribe and dates from 439 million to 408 million years before the present time. An index fossil is a fossil geologists use to date rock strata. By examining in various rock strata how these fossilized animals appeared, spread widely, and then disappeared through extinction, geologists utilize one of the surest methods of studying and dating the mammoth divisions of time they deal with.

After leaving the scenic overlook turn right down the mountain toward Bean Station. Along this section of highway, there are numerous piles of rock along the left side of the road (toward the mountainside of the road). The four-lane road construction from Thorn Hill to the bottom of the Bean Station side of the mountain began in 1976 and was completed in 1980.

During the construction, numerous landslides took place, requiring special remedial design techniques. The landslides were planar failure type slope instabilities, which means the land slides down the mountain or ridge along planes of dipping rock—usually bedding planes of the strata—into an open excavation, usually a roadway cut slope. The dipping planes of rock have been exposed along the bottom side by the excavation so as to leave the layers of rock unsupported and consequently prone to sliding. The beds of rock, where exposed near the top of the mountain, tend to dip toward the roadway.

The abundant piles of rock are actually designed embankments called rock buttresses, which are constructed to provide a resisting mass against unstable material. The rock buttresses were constructed to correct the many landslides along this stretch of the highway. In some places the sliding rock was simply blasted in place (blasted as it lay on the slope) and reshaped by bulldozers to form a stable retaining mass. This is referred to as a "shot-in-place" rock buttress. These areas are generally vegetated with trees and some grass and look like irregular cut slopes with pieces of the whitish to brownish gray sandstone protruding out of the slope.

The roadway from the bottom of the mountain over to highway U.S. 11W was completed in 1997. This project involved the removal and burial of acid-producing rock taken from the Chattanooga Shale, a black shale material containing not only pyrite (fool's gold) but also small amounts of oil and uranium. This Chattanooga shale formation can be seen near the bottom of the mountain as a blackish brown shaley material that is stained red to rusty brown in places by the acid material.

In addition, the cut slopes were covered with a black matting material and wire mesh to prevent rock fall problems and to establish vegetation on the cut slope. The planting of vegetation on the slopes has been only partially successful. At the foot of Clinch Mountain this trip turns south on 11W. Here on the way back to Knoxville, the route passes between two great ridges, and after a while the land flattens. There are snaggled-tooth gray barns and orange pumpkins in the fields.

William Bean, the first white child born in what is now Tennessee, lived and hunted here, trading shots with the Cherokees for domination, a fight the Native Americans eventually lost (Ramsey, 1853).

The ancient Indian Warpath meanders along the southern face of Clinch Mountain and then meets Boone's Trace at Bean Station, a place that became one of the most famous crossroads in the frontier's early settlement. Streams of people, stagecoaches, and drovers driving herds of livestock and fowl from Baltimore to Nashville passed through here on the tide of history. U.S. 11W is now that same road. The Carolina to Kentucky Turnpike, now the route of U.S. Highway 25E, was known as the Broadway of America, eventually unwinding into Florida.

The landscape again controls the path of the roadway as highway 11W follows Richland Valley back toward Blaine. To the right is Poor Valley Knobs, with its "cocks comb" ridgeline. Behind Poor Valley Knobs are Poor Valley and then Clinch Mountain.

Running underneath Poor Valley is a very black shale formation known as Chattanooga Shale. It is an oil shale and uranium reserve deposit and it also contains the sulfide mineral pyrite (fool's gold) which tends to produce acidic runoff when rain hits the exposed portions of the Chattanooga Shale. That's one reason why Poor Valley has very little topsoil and is very acidic. Food crops do not do well in Poor Valley. That's why early settlers called the place "Poor Valley." Very little of anything but grass can grow in the soil.

Clinch Mountain, capped by the Clinch Sandstone, looms to the right and Richland Knobs to the left as highway 11W flows to the southwest

toward Knoxville. In past times, the valley was dotted by mineral spring resorts where creeks flowed from Poor Valley into Richland Valley.

The old Tate Springs Hotel, now occupied by Kingswood Home, was one of the more famous resorts. The gazebo sits directly over the springs. Watch for it on the right as you pass by on 11W. Avondale Springs and Lee Springs had hotels and a thriving resort business in the early 1900s. Old-timers claimed that there were more mineral springs coming from Clinch Mountain than anywhere else in the world.

Capt. Thomas Tomlinson, who arrived in Grainger County and Bean Station after the close of the Civil War, started Tate Springs in what would become the world-famous Tate Springs Hotel (Brown, 1990). People came from all over, even from abroad, to stay at the captain's grand two-hundred-room hotel on Mount Cecilia. The hotel looked like a beautiful and luxurious steamboat with stacks of levels and glitter. The three-story hotel was a shining glory defined by its hand-carved balustrades along boardwalks.

The rich and the famous came calling. They took their "hot Tates" (hot mineral water from Tate Springs, brought to their rooms) in the early morning and played golf during the day and a game of tennis in the cool of the evening.

Captain Tomlinson bottled every drop of water he could and sold it across the world. That same mineral springs now empties into Cherokee Lake. The water was so good, the captain sold it for ten cents a drink in drugstores. He sold it by the jar and in one-gallon and five-gallon demijohns.

Some famous families, like the Studebakers from South Bend, Indiana, the notable automobile manufacturers, had their own suite of rooms in the captain's hotel.

The old Peavine, a railroad line running from Knoxville to Bristol, dropped off passengers at Bean Station by the hundreds. But when modern roads cut through, the masses no longer came to Tate Springs for month-long stays. After that, the hotel was utilized for overnight stays.

Soon the hotel slipped into decline, and it burned in the 1940s. Only the skeletal outlines of its former splendor can be seen now on the site of the private Kingswood School for underprivileged children. The two-story gazebo hovering over Tate Springs is the most prominent reminder of the hotel's famous and colorful past.

Instead of water, Grainger County today is celebrated for its tomatoes. The tomatoes grow large, sweet, and juicy in the county's wonderful soil, a soil derived from the limestone that underlies the area where the tomatoes are grown. It is an orange-red to reddish brown silty clay soil with siliceous chert fragments scattered throughout.

Fig. 32.
The Presbyterian Church in Rutledge was constructed around 1900 and typifies the cultural history of the East Tennessee region.

Leaving Bean Station, Clinch Mountain abruptly ends between Blaine and Luttrell where the inactive Saltville Fault, a thrust fault, slices up against the Poor Valley Knobs and extends from Knoxville into western Virginia. On the way to Blaine you pass through historic Rutledge, the county seat of Grainger County (fig. 32).

The southern end of Clinch Mountain is Signal Point. At 2,200 feet, it is the highest pointed peak that can be seen from 11W in Blaine.

In Knox County, highway 11W soon passes the large but short and stocky House Mountain. This is another one of those mountains formed by the resistant strata of the Clinch Sandstone, the cap of the mountain. The state of Tennessee has established a day-use State Natural Area on the top of House Mountain. This makes a great side trip for a good hike with astounding views of the valley. On clear days the Knoxville skyline and the Smoky Mountains can be seen from the top of House Mountain.

Color abounds everywhere in East Tennessee during October. The back roads beckon travelers and the glow of autumn sweetens the countryside.

PART III

The Cumberland Plateau

Canyon Country

Destination: *Honey Creek Overlook in the Big South Fork National Recreation Area.*

Route: *Begins and ends in downtown Knoxville; I-75, S.R. 63, U.S. 27, S.R. 52, county roads, and park roads (Burnt Mill Ford Road, Honey Creek Loop Road, and Old Mt. Helen Road). Includes Scott, Fentress, and Morgan Counties.*

Cities and towns: *Huntsville, Elgin, Rugby, Wartburg, Oliver Springs, and Oak Ridge.*

Trip length: *Approximately 180 miles roundtrip.*

Nature of the roads: *Four-lane interstate, two-lane paved state highways, paved and gravel county and park roads; some gravel roads are one-and-half lanes and very curvy and hilly–slow speeds are required.*

Special features: *Honey Creek Overlook into canyon of Big South Fork; Cumberland Plateau landscape, scenic drive through heavily forested Big South Fork Park.*

The geology of Canyon Country affords the visitor a dramatic view of the mountain-building episodes in the early life of the Alleghenies and the creation of both the Valley and Ridge physiographic province and the massive Cumberland Plateau. Along this trip, there are land surfaces of alternating valleys and ridges that ripple into mountainous topography and then drop into deep, narrow canyons. This route offers dozens of opportunities to look down into deeply incised canyons and wild rivers along both interstate highways and, at the opposite extreme, to venture onto out-of-the-way narrow gravel roads that are sometimes impassable during the winter. It's a good idea to pack a picnic lunch for this trip. By the time the route reaches the canyon overlook in the Big South Fork National River and Recreation Area, a 123,000-acre National Park that cuts across Scott, Morgan, and Fentress Counties in East Tennessee, you'll want to take the time to savor a sandwich along with the beautiful scenery.

For this trip, take Interstate 75 north from Knoxville (map 8). About twenty-six minutes from downtown, the route crosses the Clinch River, where the Cumberland Plateau and Walden's Ridge can be viewed in tandem. Both are bold and bright, appearing blue against the distant skyline.

About forty-five miles from Knoxville, observe on the right one of East Tennessee's special phenomena: Pennsylvanian-age rock—or rock that is about 250 million years old. The vertical sandstone feature here is known locally as Devil's Racetrack. This bedrock consists of alternating layers, like a sandwich. The bread is the hard sandstone, and the softer stuff inside is the shale. Originally, when natural forces began to shove the rock strata, they were horizontal. Then, some 230 million years ago, the continents collided and this rock was turned up on its end when the Pine Mountain Thrust Fault came slicing through.

Erosion wore away the softer shale, but the much harder and resistant sandstone survived. As a result, layers of sandstone form pinnacles, giant slabs that thrust into the air like great sabers, creating one of the natural wonders of East Tennessee.

At highway 63 exit from I-75 and head in the direction of Huntsville. This is roughly forty-nine miles from Knoxville. On the way to Huntsville, pay close attention to the rock outcropping along the road. These layers are horizontal, as opposed to the upturned strata visible coming up the mountain on Interstate 75. Once the trip arrives in Huntsville, it will not be difficult to find fall color. It will be all around in this mixed hardwood forest. At this point, the route is on top of the Cumberland Plateau, a physiographic province roughly a thousand feet higher than the neighboring Valley and Ridge Province you just left. Most of the layers of rock here are generally flat lying, and consist mainly of sandstone, shale, and coal seams of Pennsylvanian age.

Along the way here there are the rock remnants of a coastal lagoon. At about fifty-six miles, as the route passes through Poteet Gap and crosses into Scott County, is a good view of the Cumberland Plateau as it dominates the skyline like some great expression of energy. Imagine the impression the sight made on the first pioneers. The plateau was a lifeline, promising the prospect of wood for shelter and warmth, fields to plant for food, and abundant sources of water.

A great swath of the park lies in Scott County, one of the most independent counties in the state. During the Civil War, when Tennessee

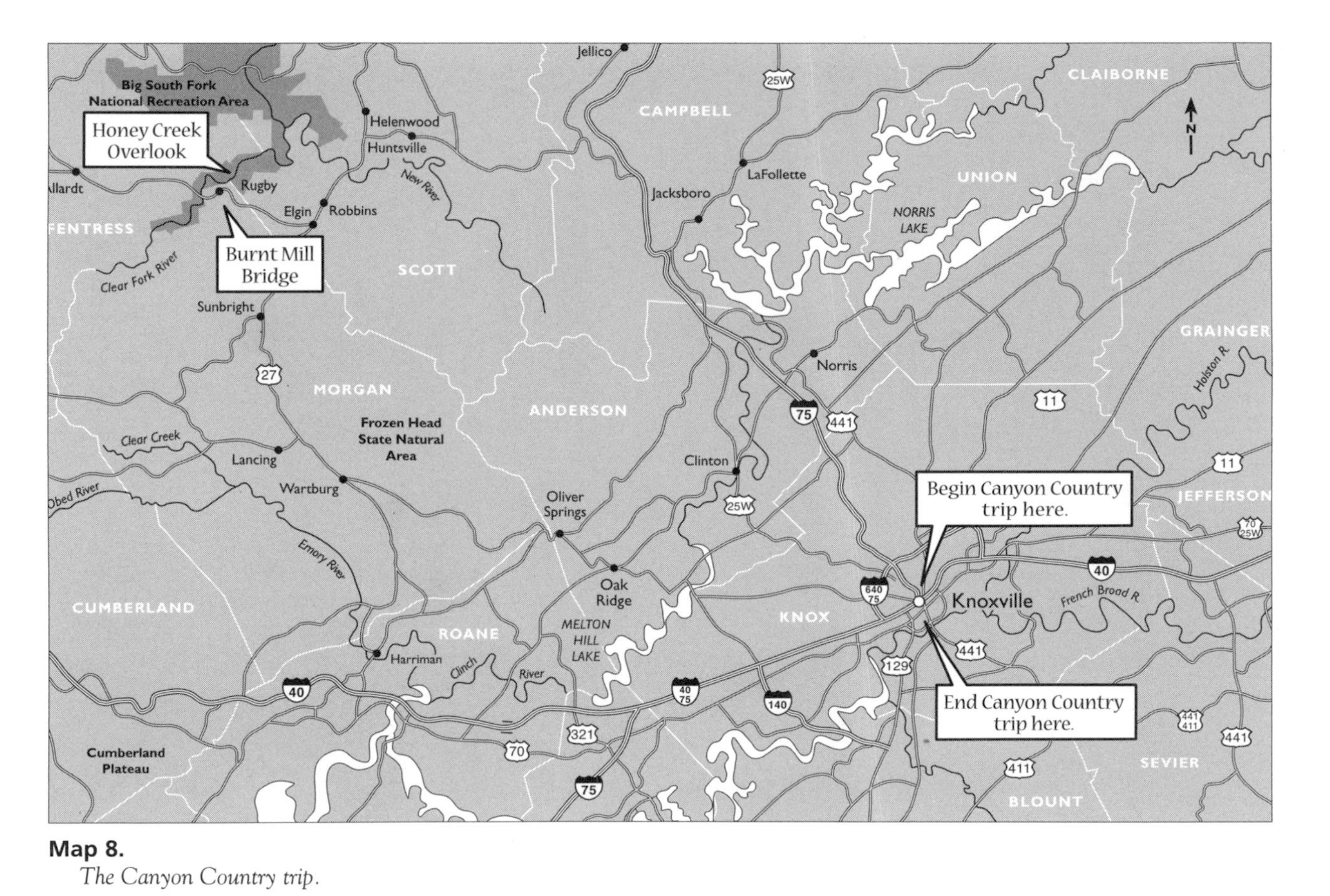

Map 8.
The Canyon Country trip.

decided to secede from the Union, Scott countians who objected to this decision themselves voted to secede from Tennessee. And they did, making Scott County the Free and Independent State of Scott. It remained independent until 1886, when the Tennessee State Legislature returned Scott County to the state's family of counties.

The Big South Fork National River and Recreation Area is situated on ancient land. Some ten thousand years ago, Woodland Indians roamed its hills and fished its streams. Here mountainous terrain is sliced with narrow valleys and carved with peaceful coves, all ablaze with color in the month of October.

The first white people to inhabit the county were long hunters from Virginia and North Carolina, who recognized the vast resources of this isolated county. Although the land is rugged and the soil less fertile than some other parts of upper East Tennessee, its early settlers were characteristically tough and independent, as are their descendants who continue to live in the county.

One of Scott County's native sons is former U.S. Sen. Howard Baker, who grew up not far from Big South Fork and was present at the ceremonies the day it was dedicated as a national recreation area. "I like this county because it still has that frontier spirit. The people here have a low threshold for tolerance and they are not impressed with names and money," the senator once said. "What I learned growing up in this county is independence, self-reliance. They make easy friendships and have an affinity for the land. I think that if William Blount or Andrew Jackson could return today, they would fit right in Scott County. This county has changed less than many others" (Brown, personal communication, c. 1985).

In the eighteenth century, Scott County constituted the private hunting grounds of Cherokee Indians and the Chickasaws, but when the indomitable Scotch-Irish pushed up against the frontier, they established "cabin rights," which meant that they owned the land where they put up a cabin and a split-rail fence.

The roadside on this trip reflects the character of the underlying rock strata. Rocks once folded like pieces of paper and their subsequent erosion have produced the Valley and Ridge topography, which is much like that found around Knoxville. Erosion of this resistant, flat-lying rock surface on the plateau usually results in relatively flat rolling lands that are dissected by deep, narrow canyons as streams eat at the rocks in a headward direction.

This trip engages both folded, upturned rocks and flat-lying strata whose ages vary from Cambrian and Ordovician (400–500 million years old) to Pennsylvanian (250 million years old). Shale and limestone are commonly found in the older strata of the Valley and Ridge, while sandstone, siltstone, carbonaceous shale, and coal seams are common in the plateau country.

Across a great split in the earth at the Honey Creek Overlook in the Big South Fork River and National Recreation Area—about one-half mile wide and six hundred feet deep—a sheer sandstone bluff with signature bulges and fissures looks as if nature has deposited its brightest colors on the maples, blackgum, sweetgum, sourwood, oaks, walnuts, and dogwoods growing along its surface. The Big South Fork River is the blue streak at the base of the gorge in the bottom of the cut, painting a startling contrast to the brilliant flame-colored leaves.

Here, time and space have merged in a vast area so huge, it is hard to comprehend the scenery in one viewing. It has to be absorbed in snippets. The hard sandstone rim of the gorge appears to be decked in color, the quintessential signature of fall in East Tennessee.

Huge and powerful forces of nature have worked this area for more than 200 million years, eroding lithified sands and clays deposited along an ancient coastline edged with barrier islands and swampy lagoons that arrived and disappeared when the continents came together and raised the region from below sea level to two thousand feet above sea level. Watch closely at Burnt Mill Bridge in Big South Fork, where, preserved in cross-bedded sandstone, are the remnants of a coastal environment with sand dunes. Mountains rise to the east and a continent to the west-northwest. Giant fern trees and bull rushes over fifteen feet tall were once the dominant vegetation along this ancient coast, but today there are few vestiges of those sediments and plants along the gorge of the Big South Fork of the Cumberland River and its tributaries.

At the Honey Creek Overlook, the Cumberland River stretches deep into the Big South Fork National River and Recreation Area. This is the optimum place to see extraordinary fall color. From the dramatic view of the canyon to the colorful canvas of the canyon rim, this is fall at its best. Besides this extravagant show, there is quiet and freedom from the motorcades of leaf watchers jammed bumper-to-bumper on the Blue Ridge Parkway or in the Great Smoky Mountains National Park.

To those who have studied the state's geologic past, this part of the Cumberland is known as the dissected plateau. This means that the landscape has been deeply eroded in many places, dismembered by numerous steams and valleys—sliced, nicked, and carved to form the incised valleys that are bounded by the higher country. This section of the trip crosses several major water drainage areas marking this portion of the plateau, including Buffalo Creek and Paint Rock Creek. Both of these are on the route before it enters Huntsville.

The plateau in this area is quite rugged. Due to its topography, the land is like a roller coaster. It twists and turns, rolls up steeply, and then comes hurtling downward. Much farther south and west, the plateau surface becomes flatter and continuous, following more closely the form of the underlying rock strata.

Scott County is shaped like an arrowhead and lies on top of the plateau, which is about two thousand feet above sea level. Its hillsides are random and slope into narrow valleys and coves.

Natural resources abound on the plateau. There were abundant trees and good earth for brick making. The discovery of coal became the bane of later populations, who had to deal with the environmental fallout of reckless greed and the disintegration of the landscape.

As the trip leaves the higher elevations of the plateau and descends into the canyon of Buffalo Creek drainage, the roadside has its own road show of chicory, jo-pye weed, laurel, ironweed, and goldenrod.

Pass through Paint Rock and cross over Paint Rock Creek, and then go on through Huntsville. At 69.9 miles, turn left onto U.S. Highway 27 South toward Robbins. After crossing these creeks, notice the massive exposures of light gray to brown Pennsylvanian-age rocks along the bluffs overlooking the streams. These rocks are mostly sandstone and siltstone that are lying flat. They were originally deposited as coastal sandbars, barrier islands, and channel sands (sand that is deposited in stream channels flowing near and into the sea). Migrating sands from dunes, bars, or beach deposits, or even by clays and muds from adjoining lagoons, gradually covered these sand deposits. Today's comparison would be the South Carolina–Georgia coastal environments where there are numerous barrier islands and many streams flowing into swampy estuaries and lagoons before draining into the ocean.

Cross New River at seventy-two miles from Knoxville. Three miles later, the route enters Robbins. Slow a bit here. There are a couple of other towns nearby—Glen Mary and Helenwood—the likes of which cannot be seen anywhere else in the state. These towns sprang up almost overnight in 1882, when coal was discovered. Nothing much is left of Glen Mary or Helenwood today, but just before the turn of the century, Helenwood alone had some thirty saloons.

As the great coal deposits played out, land speculators turned their attention to the timber that was left and took it out. Later, coal strip mines and natural gas and oil operations moved in with little regard for the environmental consequences. As coal disappeared, and coke ovens around Glen Mary cooled, Robbins sprang back to life with the production of brick. Some of the finest bricks in the nation came from here. Old-timers often referred to Robbins as "Brick Yard."

Continue on highway 27 pass Barton's Chapel. For the hungry, the G&K Grocery features ham-and-cheese sandwiches over two inches thick.

About seven miles from Robbins, you come to the community of Elgin. Here, turn right onto highway 52 West, heading toward Pickett State Park.

The next few turns can be tricky, but by following these instructions, this leg of the trip will wind up at Burnt Mill Bridge, an old steel-through-truss bridge in the Big South Fork National River and Recreation Area. Now a natural area, the Big South Fork was once the treasure chest for coal barons.

After turning right onto highway 52 West, the route will pass onto a country road about one mile from highway 27. A sign points to Concord Separate Baptist Church. Notice the old farmsteads and barns. This narrow country road is full of color, but has few automobiles. Continue for about three or four miles to the intersection of Black Creek Crossroads. On the right is Black Creek Crossroads Missionary Baptist Church.

Follow the signs straight across the road to Burnt Mill Bridge. This is a macadam road that will soon change to gravel. Macadam is an old method of paving roads by applying crushed stone to a bituminous (coal) binder. The stones stick together, creating the base of a roadway. This process results in a hard, but somewhat smooth surface, much better than mud or dirt roads, but rougher than modern-day asphalt roads. Many old county gravel roads were converted to macadam and then to some version of asphalt pavement as roadways evolved over time.

At eighty-two miles from Knoxville, turn left onto a gravel road, which goes down a steep grade. Overhead now is a thick forest of color. Just after coming down the grade, about four-tenths of a mile from the turnoff onto the gravel road, the rusting hulk of Burnt Mill Bridge is revealed. This bridge, about 180 feet across, spans Clear Fork River. Colors bounce off the water in every direction, creating a mirror image of the burnished trees.

Here the river runs around islands of green grass. The rocks beneath the river were once part of an ancient tidal channel. The rocks just under the water's surface have risen more than two thousand feet as a result of the colliding continents millions of years ago. The landscape here was once a deposit along a coastline. Think of the South Carolina beaches, and you have an idea of what it was like here in Pennsylvanian time. During the Pennsylvanian Period, the Appalachians were high, but they were no longer alpine in nature. Fern trees grew from sixty to seventy feet tall and three to four feet in diameter; bulrushes were six to ten inches in diameter and fifteen to twenty feet tall and flourished here as well. Giant dragonflies with wingspans of three feet and cockroaches were also present in this environment millions of years ago.

Reptiles also first appeared during this time, but were only about a foot long or so, and land snails also made their transition from the oceans onto dry land. Cockroaches were so abundant that the Pennsylvanian Period is sometimes referred to as the "Age of Cockroaches." Amphibians were also common, and some were over fifteen feet in length.

While crossing Burnt Mill Bridge, look back to the right to see a low rock bluff, some twenty feet high. This is an exposure of quartz sandstone, which is very hard and resistant to weathering. The lines in the rock that sweep from one side of the rock surface to the other are referred to as cross bedding by geologists. Cross bedding shows the preservation of the layering structures created as sand is moved across a surface by wind or water. Sand dunes, sandbars, channel-deposited sand, and even the ripple marks along Clear Fork River are etched into the cross-bedding structures, representing the coastal environment where these rocks were formed.

After crossing the bridge, turn up the gravel road for some of the best fall color to be found in this area. The trees here are about as small as the ones in the Great Smoky Mountains National Park were fifty years ago.

At eighty-six miles from Knoxville, turn right into the Honey Creek Trailhead Overlook—one mile from the turnoff. The short walk is well worth the view (figs. 33 and 34).

This is tabletop land, one of the most spectacular views in East Tennessee. Across the gorge is a massive sandstone cap. This is a great place to spend some time, drinking in this view.

It is worth pondering what this area might have looked like some 230 million years ago. Like those at Clear Fork River, the largely flat and massive sandstone strata exposed along the bluffs across the canyon were once sand deposits along an ancient coastline. In this ancient environment, occasional storms (not unlike the tropical variety we experience today) would churn through this area, causing much turmoil in the landscape. A close look at some of the rocks will show shredded plant debris, torn loose from the earth by the storms' thrashing winds and surf. In other rock deposits impressions of fern fronds and bulrushes preserved in the ancient sediment can be found.

Fig. 33.
This view of the canyon in the Big South Fork is at the Honey Creek Overlook. Fall colors are spectacular from this vantage point.

Cumberland Plateau

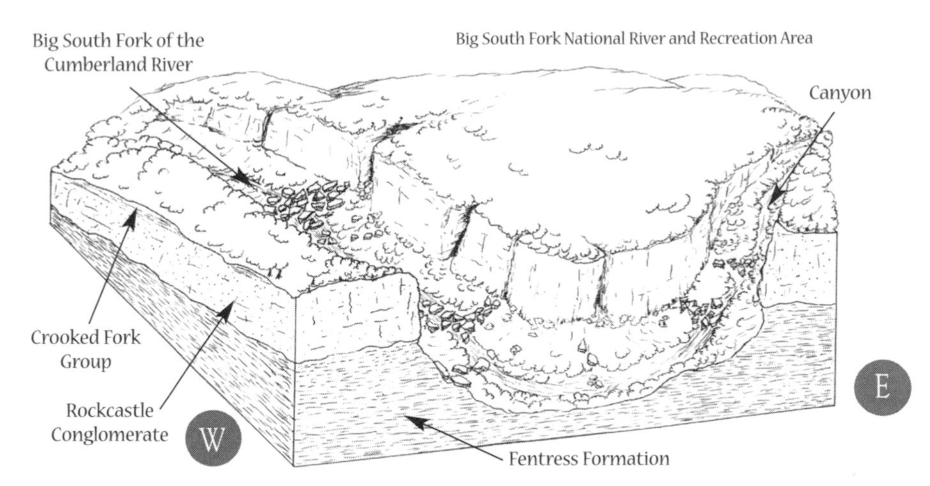

Fig. 34.
This schematic diagram shows the underlying geologic structure of the Big South Fork National River and Recreation Area. Note that the rock layers are mostly horizontal.

Fig. 35.
This old store on the outskirts of the Big South Fork area tells of the past activity in this now rural and sparsely populated plateau region.

With evergreens acting as a backdrop and the river below, this view presents color saturation at its fullest. Within view there are massive, house-sized boulders that dot the river's path. They are pieces of sandstone that once existed across the now open canyon. Over the millennia, stream erosion below cut a gash into the flat-lying rocks, forming the canyon.

After leaving the overlook parking lot, turn right onto Burnt Mill Ford Road. At about eighty-nine miles, the route leads to Old Mt. Helen Road and Honey Creek Loop Road. Turn right to leave the gravel road and proceed to Mt. Helen, on the edge of Big South Fork National River and Recreational Area and bordering Scott and Fentress Counties.

One of the next treats will be "the store" (fig. 35), an old establishment no longer in business, but it looks like a set right out of an old movie. More than a century old, the store once served as Hickory Ridge School. Local folklore says that the store was built by the community of Mt. Helen.

The trip from Burnt Mill Bridge (fig. 36) to Mt. Helen traverses some beautiful country by gravel road (fig. 37). This is a fine place to negotiate some curvy and sometimes bumpy roads and enjoy the grand but fleeting colors of autumn.

Fig. 36.
Burnt Mill Bridge crosses Clear Fork River in a small canyon of the Big South Fork area.

Fig. 37.
Gravel roads are curvy and slow in the plateau region but offer excellent views of the fall color display.

For the return trip to Knoxville, stay on Mt. Helen Road until the junction with highway 52, and then turn left. The route passes through Rugby and back to Elgin.

At Elgin, turn right, (south) on highway 27. The next turn is onto highway 62 in Wartburg. This route will lead through Oliver Springs and back to Knoxville.

Glossary

Acid mine drainage. Acidic ground and/or surface water emitted from a mine site, usually containing a percentage of sulfuric acid. Also associated with road excavation in rock containing the mineral pyrite.

Age. In geologic terms, any great period of time in the history of the earth or the material universe marked by special phases of physical conditions or organic development.

Alleghenies. A mountain range of the Appalachian system, extending from Pennsylvania to Virginia.

Alluvium. Sediments formed by rivers and streams.

Anthocyanin. The water-soluble coloring matter of flowers, leaves, and other parts of plants that imparts red, violet, or blue colors.

Anticline. A fold or arch of stratified rocks in which the strata dip in opposite directions from a common ride or axis.

Appalachians. A mountain system of eastern North America extending from the Gaspe Peninsula to Alabama; the highest peak is Mount Mitchell, N.C., 6,684 feet.

Argillite. Dark, fine-grained rock without cleavage or schistosity; resulting from low-grade metamorphism of claystone or mudstone.

Asphalt. A brown to black solid or semi-solid bituminous substance, occurring in nature but also obtained as the residue from the refining of certain petroleums and then know as artificial asphalt.

Avalanche. A large mass of snow or ice, sometimes accompanied by other material, moving rapidly down a mountain slope; sometimes applied to rapidly moving landslide debris.

Basement. An older rock mass, usually igneous or metamorphic, on which younger rocks have been deposited.

Bedding. Layering in sedimentary rocks.

Bedding plane fault. A fault surface that is parallel to the bedding plane of the constituent rocks.

Bedrock. Solid rock underlying weathered or transported material; sometimes exposed at the surface.

Bench. A level or gently sloping area interrupting an otherwise steep slope; commonly seen along highway cut slopes.

Block diagram. A three-dimensional perspective representation of geologic or topographic features, showing a surface area and generally two vertical cross-sections.

Block stream. An accumulation of rock fragments (usually boulder size or larger) that is elongated in the slope direction; lengths of the boulder stream can range from thirty feet to over three thousand feet; sometimes referred to as colluvium.

Boulder. A large, rounded block of stone lying on the surface of the ground or sometimes embedded in loose soil.

Brachiopod. A marine shellfish with two bilaterally symmetrical shells.

Buttress. A structure of masonry, wood, or quarried rock that gives support or stability to a slope or other structure.

Calcareous. Containing calcium carbonate.

Calcite. A mineral, calcium carbonate ($CaCO_3$), the principal constituent of limestone.

Cambrian. The first period of the Paleozoic era, from 500 to 600 million years ago.

Carbonaceous. Containing or composed of carbon; pertaining to sediment containing organic matter.

Carbonate. Containing carbon and oxygen in combination with sodium, calcium, or other elements, particularly in limestone or dolomite.

Carotin. A deep yellow or red crystalline hydrocarbon that acts as a plant pigment, especially in carrots, and occurs also in various animal tissues, where it is changed to vitamin A; also spelled "carotene."

Cave. A natural cavity, recess, chamber, or series of chambers beneath the surface of the earth, generally produced by the solution of limestone.

Cedar glade. A local vegetation dominated by cedar trees, usually found on shallow limestone soils.

Cenozoic. The most recent of the four eras into which geologic time is divided, extending from the close of the Mesozoic era to and including the present; it includes the periods called Tertiary and Quaternary in the United States.

Chalcopyrite. A widely occurring ore of copper that also contains iron and a sulfur compound known as fool's gold.

Chert. Also called flint, a hard, extremely dense sedimentary rock consisting predominantly of cryptocrystalline silica; a variety of the mineral quartz.

Chlorophyll. The green photosynthetic coloring matter of plants found in chloroplasts and made up chiefly of a blue-black ester.

Chloroplast. A plastid containing chlorophyll that is the seat of photosynthesis and starch formation.

Cleavage. The tendency for rocks to split along definite planes that generally have no relation to bedding.

Colluvium. Loose and incoherent deposits, usually at the foot of a slope or cliff and brought there chiefly by gravity; also referred to as block stream.

Conglomerate. Rock composed of rounded, waterworn fragments of older rock, usually in combination with sand.

Continental rocks. Rocks making up the continents, lighter in color and density than those formed in ocean basins.

Cross-bedded. Deposited at an angle to the horizontal; also called current bedding.

Cross-section. A profile portraying an interpretation of a vertical section of the earth explored by geological and/or geophysical methods; a short profile section usually perpendicular in orientation to a profile or survey line.

Crust. The outer most layer or shell of the earth.

Crystalline. Composed wholly of crystal mineral grains; that is igneous and metamorphic rocks as distinct from sedimentary rocks.

Cut. An excavation of soil and/or rock in a hill, ridge, or mountain, where the result is an artificially smooth or sculptured slope in the earth.

Debris. Rock and mineral fragments produced by weathering of rocks; synonymous with detritus.

Debris flow. A general designation for all types of rapid movement involving rock and mineral fragments of various kinds and conditions.

Deciduous. Falling off or shedding leaves seasonally, and leafless for part of the year.

Deformation. Change of shape or attitude of a rock body by folding, shearing, fracturing, compression, etc.

Devonian. The fourth period of the Paleozoic era, from 350 to 400 million years ago

Dip. The angle at which a stratum or any planar feature is inclined from the horizontal.

Discontinuity. A sudden change in the character of rock, such as joints, cleavage, bedding, faults, or contacts between similar or dissimilar materials.

Dolomite. A rock composed essentially of the mineral of the same name, $(CaMg)CO_3$.

Dolostone. A calcareous sedimentary rock composed of the mineral dolomite; sometimes used for the rock dolomite.

Elliptic. Pertaining to, or shaped like, an ellipse; oblong with rounded ends.

Embankment. An artificial accumulation of soil and rock placed by mechanical means in order to build a roadway; used to fill across valleys.

Equinox. Either of the two times each year when the sun crosses the equator and day and night are everywhere of equal length, being about March 21 and September 23.

Escarpment. A cliff or steep slope edging a region of higher land.

Fault. A fracture in the earth's crust along which rock on one side of the fracture has been displaced relative to rock on the other.

Fault scarp. A cliff formed by a fault, usually modified by erosion.

Feldspar. A group of abundant light-colored, rock-forming minerals belonging to the silicate class.

Fill. The placement by mechanical means of soil and rock to form a mound or raised earth structure.

Floodplain. The surface of relatively smooth land adjacent to a river channel, constructed by the present river and covered with water when the river overflows its banks at times of high water.

Fluvial. Of or pertaining to rivers; produced by river action.

Fold. A curve or bend of the rock strata, usually the result of deformation.

Foldbelt. A linear region that has been subject to folding and other deformation during mountain building.

Foliation. The parallel alignment of platy mineral grains or flattened aggregates in a metamorphosed or sheared rock.

Footwall. The underlying side of a fault.

Fossil. The remains or traces of a plant or animal, preserved in rock.

Formation. A distinctive body of one kind or related kinds of rocks, selected from a succession of strata for convenience in mapping, description, and reference.

Garnet. A silicate mineral used as a gem and as an abrasive.

Geomorphology. A branch of geology that deals with the earth's surface features and landforms.

Gneiss. A visibly crystalline metamorphic rock possessing mineral layering or foliation but not easily split along foliation surfaces.

Granite. A visibly grained igneous rock composed essentially of alkali feldspars and quartz; commonly any rock of this composition and texture, weather igneous or metamorphic in origin.

Granitic. Pertaining to granite, or similar to granite in composition or texture.

Groundwater. Subsurface water filling rock pore spaces, cracks, or solution channels.

Group. In geologic terms, a stratagraphic unit consisting of several formations, usually originally a single formation subdivided by subsequent research.

Hematite. A common iron mineral (Fe_2O_3).

Hogback. A ridge composed of a resistant layer within steeply tilted, eroded strata.

Igneous. A term applied to rocks formed by crystallization or solidification from natural silicate melts, generally at temperatures between 600 degrees C and 1000 degrees C.

Index fossil. Guide fossil; a fossil characteristic of an assemblage of fossil organisms and so far as known restricted to it.

Interglacial. The time between major advances of continental glaciers.

Intrusive. Having penetrated, while fluid, into or between other rocks but solidifying before reaching the surface.

Invertebrates. Animals without backbones.

Joint. A fracture in a rock along which no appreciable movement has occurred.

Karst. A distinctive type of landscape in which solution in limestone layers has caused abundant caves, sinkholes, or solution valleys, often with red soil residue

Limestone. A sedimentary rock composed of calcite, the mineral form of calcium carbonate; loosely, either limestone or dolomite, a very similar rock that includes considerable magnesium in its chemical composition.

Lithology. The description of rocks on the basis of such characteristics as color, mineralogical composition, structures, and grain size.

Lithosphere. The earth's crust; the outer most portion or shell of the globe, as distinguished from the underlying barysphere or centrosphere.

Long hunter. Early pioneer who hunted long distances from home over long periods of time.

Macadam. An early road made of small stone layers.

Magma. A hot mobile silicate mixture of crystals and melt within the earth's crust.

Mantle. The zone of the earth below the crust and above the core (to a depth of 3,480 kilometers).

Marble. Recrystallized limestone or dolomite; a metamorphic rock.

Mass wasting. A variety of processes by which large masses of earth material are moved by gravity, either quickly or slowly, from one place to another.

Meander. One of a series of somewhat regular and looplike bends in the course of a stream, developed when the stream is flowing at grade, through lateral shifting of its course toward the convex sides of the original curves.

Mesozoic. One of the grand divisions or eras of geologic time, following the Paleozoic and succeeded by the Cenozoic era, and comprising the Triassic, Jurassic, and Cretaceous periods.

Metamorphic rock. Any rock that has been recrystallized at a red heat, deep within the earth's crust; usually contains mineral crystals large enough to see and has a streaked or grainy appearance.

Metamorphism. The process whereby sedimentary or igneous rocks have been altered by heat and pressure accompanying deep burial in the earth's crust.

Metaquartzite. A quartzite formed by metamorphic recrystallization.

Mineral. A homogeneous, naturally occurring, solid substance of inorganic composition, consistent physical properties, and specified chemical composition.

Mississippian. The fifth period of the Paleozoic Era, from 310 to 350 million years ago.

Mollusk. An invertebrate belonging to the phylum Mollusca, characterized by a nonsegmented body that is bilaterally symmetrical and by a radially or biradially symmetrical mantle and shell.

Morphology. The observation of the form of lands.

Mudflow. A flowage of heterogeneous debris lubricated with a large amount of water, usually following a former stream course.

Nectar. The saccharine secretion of plants, collected by bees to make honey; in Greek mythology, the drink of the gods.

Oceanic rocks. Rocks making up the oceanic crust, usually darker and denser than those of the continents.

Ordovician. The second period of the Paleozoic Era, from 430 to 500 million years ago.

Orogeny. The process of the formation of mountains.

Outcrop. Any exposure of bedrock.

Overthrust. A type of fault in which an extensive slab of rock is moved across a nearly horizontal surface.

Oxbow lake. A crescent-shaped body of water occupying the abandoned meander of a stream channel.

Paleozoic. The first era of Phanerozoic geologic time, from 225 to 600 million years ago.

Phanerozoic. Comprises Paleozoic, Mesozoic, and Cenozoic; eon of evident life.

Photosynthesis. Synthesis of chemical compounds with the aid of radiant energy and especially light.

Phyllite. A metamorphic rock similar to schist but with grains so fine that they cannot be seen with the unaided eye.

Phylum. A major unit in the taxonomy of animals, ranking above "class" and below "kingdom."

Planar failure. A landslide type (also called a block glide) in which movement of material occurs along a flat, inclined plane, usually along and parallel to bedding layers of rock.

Plateau. Any comparatively flat area of great extent and elevation.

Plate tectonics. The movement of large blocks of the earth's crust over the surface of the globe.

Precambrian. Geologic time before the Paleozoic Era.

Pre-split face. In highway cutslopes, a smooth rock face developed by drilling closely spaced holes and lightly blasting to form a crack in the rock that connects the drilled holes.

Pyrite. A metallic, brass-colored iron-ore mineral (FeS_2), often called "fool's gold," used as a source of sulfur for sulfuric acid.

Quartz. A hard, glassy mineral, silicon dioxide (SiO_2), that is one of the most common rock-forming minerals, and a silicate group mineral.

Quartzite. A sedimentary or metamorphic rock composed largely of quartz grains cemented by silica.

Quaternary. The younger of the two geologic periods in the Cenozoic Era; Includes all geologic time and deposits from the end of the Tertiary period until and including the present; subdivided into the Pleistocene and Recent epochs or series.

Recrystallization. Alteration of rocks whereby pre-existing mineral grains are destroyed and new ones are formed, generally by increased heat and pressure; one of the metamorphic processes.

Reef. A range or ridge of rocks lying at or near the surface of water; either moundlike or layered, built by sedimentary organisms such as corals, and usually enclosed in rock of a differing lithology.

Regolith. The layer or mantle of loose, incoherent rock material, of whatever origin, that nearly everywhere forms the surface of the land and rests on the hard bedrocks.

Relief. The actual physical shape, configuration, or general unevenness of the earth's surface; the difference in elevation between the high and low points of a land surface.

Residuum. Soil formed in place by the disintegration and decomposition of rocks and the consequent weathering of the mineral materials.

Sandstone. A sedimentary rock composed of sand grains.

Saprolite. A soft, earthy, clay-rich, thoroughly decomposed rock formed in place by chemical weathering of igneous or metamorphic rocks.

Scarp. Cliff or steep break in a slope.

Schist. A visibly crystalline metamorphic rock containing abundant mica or other cleavable minerals so aligned that the rock breaks regularly along the mineral grains.

Sedimentary. Formed of fragments of other rocks transported from their source and deposited in water; also, transported in solution and deposited by chemical or organic agents.

Serrate. Toothed or notched like a saw.

Shale. Platy sedimentary rock formed from mud or clay, breaking easily parallel to the bedding.

Silica. Silicon dioxide (SiO_2), occurring as quartz and as a major part of many other minerals.

Siliceous. Containing abundant silica, especially as free silica rather than as silicates.

Siltation. The deposition by a stream or a river of graded particulates (sand, silt, clay) that tend to cover a stream bottom.

Siltstone. A sedimentary rock composed of mostly silt-sized particles (1/256 to 1/16 mm in diameter).

Sinkhole. A large depression caused by collapse of the ground into an underlying limestone cavern.

Slide scar. A term used to describe the resulting surface following a landslide.

Solstice. The time of year when the sun is at its greatest distance from the celestial equator and seems to pause before returning on its course; either summer solstice, about June 22 in northern hemisphere, or winter solstice, about December 22.

Strata. Beds or layers of sedimentary rock.

Stratified. Deposited in nearly horizontal layers of strata on the earth's surface.

Stratigraphic. Pertaining to the composition, sequence, and correlation of stratified rocks.

Strike. The direction or bearing of a horizontal line on a sloping bed, fault, or other rock surface.

Subsidence. A sinking movement of soil and/or rock, in which there is no free side and surface material is displaced vertically downward with little or no horizontal component.

Syncline. A fold in stratified rocks in which the strata on opposite sides usually dip inward toward each other.

Talus. Blocks of rock pried loose by frost wedging; a type of colluvium, usually found at the base of a cliff or other steep slope.

Tectonic. Pertaining to the larger structural features of the earth's crust and the forces that have produced them.

Terrace deposit. A deposit of alluvial origin, composed of pebbles and cobbles in a sand or silt matrix, which is found on a relatively flat, elongated surface along the side of a valley; the deposit is formerly the alluvial floor of the valley.

Tertiary. The earlier of the two geologic periods making up the Cenozoic Era

Thrust fault. A fault, commonly of low dip, on which rocks have slid or have been pushed laterally over other rocks.

Topography. The physical features of a district or region as represented on maps; the relief, contour, or shape of the land surface.

Unconformity. A buried surface of erosion or nondeposition separating younger rocks from younger ones.

Water gap. A pass in a ridge that is still occupied by the stream that formed the gap.

Weathering. The processes that cause solid rock to decay into soil.

Wedge failure. A landslide type in which movement occurs along the line of intersection of two inclined planes of discontinuity, resulting in a wedge-shaped scar on the landscape.

Window. A hole produced by erosion through a thrust fault, exposing the underlying rocks; Cades Cove and Tuckaleechee Cove are examples.

Xanthophyll. Any of several neutral yellow carotenoid pigments, especially that are oxygen derivatives of carotenes.

References

Aycock, J. H. 1981. "Construction Problems Involving Shale in a Geologically Complex Environment—SR 32, Grainger Co., Tennessee." In *Proceedings of the 32nd Annual Highway Geology Symposium*, 36–58.

Blount County Historic Trust. 1991.

Brown, F. W. 1990. *The Faces of East Tennessee*. Knoxville: *Knoxville News-Sentinel*. 89 pp.

———. 1991. *Coker Creek: Crossroads to History*. Coker Creek Ruritan Club. 110 pp.

———. c. 1985. Personal communication with Howard H. Baker.

Byerly, D. W. 1981. "Evaluation of the Acid Drainage Potential of Certain Precambrian rocks in the Blue Ridge province." In *Proceedings of the 32nd Annual Highway Geology Symposium*, 174–85.

———. 1997. "A Field Trip Excursion Guide, Overview of the Geology." In *Proceedings of the 48th Annual Highway Geology Symposium*, 227–52.

Ferris, William. 1989. Entry for Lomax, John A. *Encyclopedia of Southern Culture*. Univ. of Miss., Center for the Study of Southern Culture, 1069–70.

Goodspeed's History of Tennessee. Reprinted from Goodspeed's History of Tennessee. 1887. The Goodspeed Publishing Co., 1,317 pp.

Hatcher, R. D., Jr. 1989. "Tectonic Synthesis of the U.S. Appalachians." In *The Appalachian–Ouachita Orogen in the United States*. Ed. R. D. Hatcher Jr., W. A. Thomas, and G. W. Viele. Boulder, Geological Society of America, The Geology of North America, vol. F–2, 511–36.

Klepser, H. J. 1967. Historical Geology Laboratory Manual. Department of Geology, Univ. of Tenn., Knoxville. 85 pp.

Lindsey, Anne H., and C. Ritchie Bell. 1983. *Fall Color and Woodland Harvests*, Laurel Hill Press, 184 pp.

Little, Elbert J. 1979. *Checklist of United States Trees*. Agriculture Handbook No. 541. USDA Forest Service. 375 pp.

Moore, H. L. 1986. *A History of the Churches of Grainger County, Tennessee*. Grainger County Historical Society, 143 pp.

———. 1997. "Construction Techniques on U.S. 23 (Future I-26), Unicoi County, Tennessee." In *Proceedings of the 48th Annual Highway Geology Symposium*, ed. D. W. Byerly, 253–73.

Oriel, S. S. 1951. "Structure of the Hot Springs Window, Madison County. North Carolina." *American Journal of Science* 275-A: 298–336.

Price, Henry R. 1971. "Melungeons: The Vanishing Colony of Newman's Ridge," Pamphlet Presented at the Spring Meeting of the American Studies Association of Kentucky and Tennessee, Tennessee Technological University, Cookeville, Tenn., March 25–26. 26 pp.

Radford, A. E., H. E. Ahles, and C. R. Bell. 1983. *Manual of the Vascular Flora of the Carolinas*. Chapel Hill: Univ. of N.C. Press. 1,183 pp.

Ramsey, Dr. J. G. M. *The Annals of Tennessee to the End of the Eighteenth Century*. Originally printed in 1853 for J. G. M. Ramsey. Charleston, S.C.: Walker and Jones. Reprinted, 1967. 821 pp.

Tarbuck, E. J., and F. K. Lutgens. 1996. *Earth: An Introduction to Physical Geology*. Upper Saddle River, N.J.: Prentice Hall. 605 pp.

West, Carroll Van, et al., eds. *Tennessee Encyclopedia of History and Culture*. Nashville: Tennessee Historical Society, 1998.

Wilson, R. L. 1981. Guide to the Geology along the Interstate Highways in Tennessee. Tenn. Div. Geol. Rep. Invest. 39. 79 pp.

Index

References to illustrations are printed in **boldface**.